Separation Techniques and Circular Economy

Separation Techniques and Circular Economy

Da-Qi Cao

MDPI

Basel • Beijing • Wuhan • Barcelona • Belgrade • Novi Sad • Cluj • Manchester

Editor
Da-Qi Cao
Sino-Dutch R&D Centre for
Future Wastewater Treatment
Technologies
Beijing University of Civil
Engineering and Architecture
Beijing
China

Editorial Office
MDPI AG
Grosspeteranlage 5
4052 Basel, Switzerland

This is a reprint of articles from the Special Issue published online in the open access journal *Membranes* (ISSN 2077-0375) (available at: www.mdpi.com/journal/membranes/special_issues/sep_tech_circular_econ).

For citation purposes, cite each article independently as indicated on the article page online and as indicated below:

Lastname, A.A.; Lastname, B.B. Article Title. *Journal Name* **Year**, *Volume Number*, Page Range.

ISBN 978-3-7258-2430-4 (Hbk)
ISBN 978-3-7258-2429-8 (PDF)
doi.org/10.3390/books978-3-7258-2429-8

Cover image courtesy of Da-Qi Cao

Contents

About the Editor

Da-Qi Cao

Da-Qi Cao received his Ph.D. in Chemical Engineering and Biotechnology from Nagoya University, Japan. He is an Associate Professor at Beijing University of Civil Engineering and Architecture and a Visiting Scholar at Yau Mathematical Sciences Center, Tsinghua University, and Michigan State University. He has long been engaged in cutting-edge research of environmental protection, including wastewater resource utilization, novel membrane materials, trace antibiotic control, intelligent water management, and carbon neutral technology. He has obtained awards for Young Talents of Beijing Municipal Institutions, Beijing Nova Program, and Beijing Outstanding Talents, and he has presided over seven projects at the provincial level or above, as well as published more than 50 academic papers (20 first-author SCI papers, of which 12 are JCR Q1, and 1 paper was awarded the Outstanding Paper Award by The Society of Chemical Engineers, Japan), an academic monograph "Recycling of polymeric substances from sewage" (260,000 words), and one chapter of English academic monograph. He has applied for 23 invention patents (11 authorized) and is an expert in reviewing more than 10 international renowned journals.

Preface

Efficient separation techniques play an important role in the process of resource recovery, and these techniques include physical, chemical, physicochemical, and/or biological methods that are selected for their low cost and low energy consumption and for being free of secondary pollution. Additionally, the highest possible value added of the separated products is obtained to enhance the economy. This reprint presents novel separation technologies involving the recovery of extracellular polymeric substances, calcium alginate in wastewater treatment, plasmid DNA in gene therapy, lactose in the dairy industry, lactic acid in the fermented broth of kitchen waste, phenylalanine losses in neutralization dialysis, the removal of secondary effluent organic matter, acetaldehyde in the atmosphere, and the purification of the liquid fraction of digestates. We believe that the contributions of this research effort will benefit resource recovery.

Da-Qi Cao

Editor

membranes

Editorial

Separation Techniques and Circular Economy

Da-Qi Cao [1,2]

1 Sino-Dutch R&D Centre for Future Wastewater Treatment Technologies/Key Laboratory of Urban Stormwater System and Water Environment, Beijing University of Civil Engineering and Architecture, Beijing 100044, China; caodaqi18@163.com; Tel.: +86-(0)-10-6832-2123; Fax: +86-(0)-10-6832-2128

2 Yau Mathematical Sciences Center, Tsinghua University, Beijing 100084, China

Efficient separation techniques play an important role in the process of resource recovery, and these techniques include physical, chemical, physicochemical, and/or biological methods that are selected for their low cost and low energy consumption and for being free of secondary pollution. Additionally, the highest possible value added of the separated products is obtained to enhance the economy. This current Special Issue titled "Separation Techniques and Circular Economy" in *Membranes* presents the novel separation technologies involving the recovery of extracellular polymeric substances (EPSs) [1], calcium alginate (Ca-Alg) [2] in wastewater treatment, plasmid DNA in gene therapy [3], lactose in dairy industry [4], lactic acid (LA) in the fermented broth of kitchen waste [5,6], phenylalanine losses in neutralization dialysis (ND) [7], the removal of secondary effluent organic matter (EfOM) [8], and acetaldehyde (CH_3CHO) in the atmosphere [9].

For the recycling of EPSs from excess sludge in wastewater treatment plants (WWTPs), the potential of nanofiber membranes for use in effective concentration and the recycling of EPSs via membrane separation was indicated [1]. Electrospun nanofiber membranes (ENMs) were fabricated using polyvinylidene fluoride (PVDF) and used for the recovery of EPSs extracted from the excess sludge using the cation exchange resin (CER) method. The fabricated ENM containing 14 wt.% PVDF showed excellent properties, with a high average water flux (376.8 L/(m^2·h)) and an excellent EPS recovery rate (94.1%) in the dead-end filtration of a 1.0 g/L EPS solution at 20 kPa. The ENMs displayed excellent mechanical strength, antifouling properties, and high reusability after five recycles. The filtration pressure had a negligible effect on the average EPS recovery rate and water flux. The novel dead-end filtration with an EPS filter cake on the ENM surface was effective in removing heavy-metal ions, with the removal rates of Pb^{2+}, Cu^{2+}, and Cr^{6+} being 89.5%, 73.5%, and 74.6%, respectively.

Ca-Alg is a novel target product for recovering alginate from aerobic granular sludge. A novel production method, where Ca-Alg was formed in a sodium alginate (SA) feed solution (FS) and concentrated via forward osmosis (FO) with Ca^{2+} reverse osmosis using a draw solution of $CaCl_2$, was proposed [2]. An abnormal reverse solute diffusion was observed, with the average reverse solute flux (RSF) decreasing with increasing $CaCl_2$ concentrations, while the average RSF increased with increasing alginate concentrations. The RSF of Ca^{2+} in FS decreased continuously as the FO progressed using 1.0 g/L SA as the FS, while it increased initially and later decreased using 2.0 and 3.0 g/L SA as the FS. These results were attributed to the Ca-Alg recovery production (CARP) that formed on the FO membrane's surface on the feed side, and the percentage of Ca^{2+} in CARP relative to total Ca^{2+} reverse osmosis reached 36.28%. Scanning electron microscopy and energy dispersive spectroscopy also verified CARP's existence and its Ca^{2+} content. The thin-film composite FO membrane with a supporting polysulfone electrospinning nanofiber membrane layer showed high water flux and the RSF of Ca^{2+}, which was proposed as a novel FO membrane for Ca-Alg production via the FO process with Ca^{2+} reverse diffusion. Four mechanisms, including the molecular sieve role, the electrification of colloids, the osmotic pressure of ions in CARP, and the FO membrane structure, were proposed to control Ca-Alg production.

Citation: Cao, D.-Q. Separation Techniques and Circular Economy. *Membranes* **2023**, *13*, 778. https://doi.org/10.3390/membranes13090778

Received: 25 August 2023
Accepted: 30 August 2023
Published: 4 September 2023

Plasmid DNA is used as a vector for gene therapy and DNA vaccination. Because of its circular structure, a significant deformation occurs during filtration and easily permeates the membrane, hindering the selection of separation membranes based on molecular weight. Affinity microfiltration (MF) and plasmid DNA purification were applied [3]. *A*-Fe_2O_3 with an isoelectric point of approximately 8 and a particle size of 0.5 μm was selected as the ligand for the two-stage affinity MF of plasmid DNA. In the first stage of MF, the experiment was conducted at a pH of 5, and a cake of α-Fe_2O_3 with bound plasmid DNA was obtained. Next, liquid permeation (pH 9 and 10) through the cake was performed to elute the bound plasmid DNA. Plasmid DNA was eluted during the early phase of liquid permeation at a pH of 10. Furthermore, agarose gel analysis confirmed the usefulness of the two-stage affinity MF method with the adsorption and desorption for plasmid DNA purification.

The lactose in the dairy industry is not produced profitably. Nanofiltration (NF) concentration has shown to be a good choice since partial demineralization can be realized in parallel. Ten commercial polymer NF membranes were systematically studied in detail for their suitability to concentrate lactose, with the proviso of high flux and high to complete rejection [4]. Preliminary trials were conducted using flat-sheet membranes and a lactose model solution, and the influence of transmembrane pressure, temperature, and lactose concentration was studied. Finally, the results were evaluated using spiral wound modules and real industrial whey permeate. Membrane screening is essential since no correlation between molecular weight cut-off and permeate flow could be found. The conclusions found for the lactose model solution were in good agreement with the whey permeate, but as ions contribute to the osmotic pressure of the feed, the deviations increase during the concentration process since ions are also partly retained.

Lactic acid (LA) is an important chemical material facing rapid demand in recent years. The oriented fermentation of kitchen waste is a promising route for economic LA production. However, the purification of the LA from the broth is still troublesome when considering the economy. The purification performance of MF membrane technology for a fermentation broth from kitchen waste was investigated [5]. The results indicated that the MF process could be a desirable broth purification process to some extent, and it is promising in actual application. Under the optimum pressure of 100 kPa, a pH of 6.0, and a backflushing mode with deionized water for 3 min, the best performance was achieved, with chroma removal, turbidity removal, protein removal, and total sugar removal efficiencies of 60, 92.8, 57.64, and 32.93%, respectively. On the other hand, the refinement of LA from fermentation broth is also a spiny issue. The performance of the ultrafiltration (UF) process for the refinement of LA from the pre-microfiltered broth of kitchen waste fermentation was investigated [6]. With a 50 kDa polyethersulfone membrane, under the optimum pressure of 120 kPa, the pH of 6.0, and the backflushing mode with the deionized water for 3 min, the best performance was achieved with respect to the chroma removal efficiency, turbidity removal efficiency, protein removal efficiency, and total sugar removal efficiency of 54.3%, 89.8%, 71.7% and 58.5%, respectively, and the LA recovery efficiency was 93.6%. The results indicated that the UF process could further effectively refine the pre-microfiltered broth of kitchen waste fermentation, and the combination of MF and UF process is ideal for achieving desirable LA refinement performance.

Amino acid (phenylalanine (Phe)) and mineral salt (NaCl) solution separation can be carried out using neutralization dialysis (ND) [7]. A non-steady state mathematical model of the Phe and NaCl solution's separation via ND in a batch mode is proposed. The model takes into account the characteristics of membranes (thickness, ion-exchange capacity, and conductivity) and solutions (concentration and composition). Compared to previously developed models, the new model considers the local equilibrium of Phe protolysis reactions in solutions and membranes and the transport of all phenylalanine forms (positively and negatively charged zwitterionic) through membranes. A series of experiments on the ND demineralization of the NaCl and Phe mixed solution was carried out. In order to minimize Phe losses, the solution pH in the desalination compartment was controlled by changing the concentrations of the solutions in the acid and alkali compartments of the ND cell. The validity of the model was verified by carrying out a comparison of simulated and experimental time dependencies with respect to the solution's electrical conductivity and pH, as well as the concentration of Na^+, Cl^- ions, and Phe species in the desalination compartment. Based on the simulation's results, the role of Phe transport mechanisms in the losses of this amino acid during ND was discussed. In the experiments carried out, the demineralization rate reached 90%, accompanied by minimal Phe losses of about 16%. Modeling predicts a steep increase in Phe losses when the demineralization rate is higher than 95%. Nevertheless, simulations show that it is possible to achieve a highly demineralized solution (by 99.9%) with Phe losses amounting to 42%.

The reuse of wastewater has been identified as an important initiative for the sustainable development of the environment; thus, the removal of secondary EfOM to ensure the safety of reused wastewater is the key step and a subject of extensive research. $Al_2(SO_4)_3$ and anionic polyacrylamide were selected as the coagulant and flocculant, respectively, for the treatment of secondary effluents from a food-processing industry WWTP to meet the standard regulatory specifications for water reuse [8]. This process is efficient and economically viable for EfOM removal to realize food-processing wastewater reuse. The removal efficiencies of chemical oxygen demand (COD), components with UV_{254}, and specific ultraviolet absorbance were 44.61%, 25.13%, and 9.13%, respectively, with an associated reduction in chroma and turbidity. The fluorescence intensities (Fmax) of two humic-like components were reduced during coagulation, and the microbial humic-like components of EfOM had a better removal efficiency because of a higher $\mathrm{Log}K_\mathrm{m}$ value of 4.12. Fourier-transform infrared spectroscopy showed that $Al_2(SO_4)_3$ could remove the protein fraction of the soluble microbial products (SMPs) of EfOM by forming a loose SMP protein complex with enhanced hydrophobicity. Furthermore, flocculation reduced the aromaticity of secondary effluent. The cost of the proposed secondary effluent treatment was 0.034 CNY t^{-1} $\%COD^{-1}$.

CH_3CHO in the atmosphere is associated with adverse health effects. Among the various options for use in removing CH_3CHO, adsorption is often employed because of its convenient application and economical processes, particularly when using activated carbon. The surface of activated carbon has been modified with amines to remove CH_3CHO from the atmosphere via adsorption. However, these materials are toxic and can have harmful effects on humans when the modified activated carbon is used in air-purifier filters. Therefore, customized bead-type activated carbon (BAC) with surface modification options via amination was evaluated for removing CH_3CHO [9]. Various amounts of non-toxic piperazine or piperazine/nitric acid were used in amination. Chemical and physical analyses of the surface-modified BAC samples were performed using Brunauer–Emmett–Teller measurements, elemental analyses, and Fourier-transform infrared and X-ray photoelectron spectroscopy. The chemical structures on the surfaces of the modified BACs were analyzed in detail using X-ray absorption spectroscopy. The amine and carboxylic acid groups on the surfaces of the modified BACs are critical in CH_3CHO adsorption. Notably, piperazine amination decreased the pore size and volume of the modified BAC, but piperazine/nitric acid impregnation maintained the pore size and volume of the modified BAC. In terms of CH_3CHO adsorption, piperazine/nitric acid impregnation resulted in superior performances, with greater chemical adsorption. The linkages between the amine and carboxylic acid groups may function differently in piperazine amination and piperazine/nitric acid treatment.

Overall, the Special Issue presents helpful contributions to the recovery of biopolymers (EPS, Ca-Alg, and DNA) or micromolecules (lactose, LA, and phenylalanine) and the removal of EfOM and CH_3CHO, in which various separation techniques such as ENM separation, MF, UF, NF, FO, ND, flocculation, and adsorption are used. However, in practice, separation techniques that are involved in the circular economy include numerous methods, and these are dependent on the recycled or separated products. In conclusion, the editors appreciate the authors' and reviewers' valuable contributions to this Special Issue.

Acknowledgments: We thank the editorial board of Membranes for her invaluable support.

Conflicts of Interest: The authors declare no conflict of interest.

References

1. Cao, D.-Q.; Liu, X.-D.; Han, J.-L.; Zhang, W.-Y.; Hao, X.-D.; Iritani, E.; Katagiri, N. Recovery of Extracellular Polymeric Substances from Excess Sludge Using High-Flux Electrospun Nanofiber Membranes. *Membranes* **2023**, *13*, 74. [CrossRef]
2. Cao, D.-Q.; Tang, K.; Zhang, W.-Y.; Chang, C.; Han, J.-L.; Tian, F.; Hao, X.-D. Calcium Alginate Production through Forward Osmosis with Reverse Solute Diffusion and Mechanism Analysis. *Membranes* **2023**, *13*, 207. [CrossRef] [PubMed]
3. Katagiri, N.; Shimokawa, D.; Suzuki, T.; Kousai, M.; Iritani, E. Separation Properties of Plasmid DNA Using a Two-Stage Particle Adsorption-Microfiltration Process. *Membranes* **2023**, *13*, 168. [CrossRef] [PubMed]
4. Hofmann, K.; Hamel, C. Screening and Scale-up of Nanofiltration Membranes for Concentration of Lactose and Real Whey Permeate. *Membranes* **2023**, *13*, 173. [CrossRef] [PubMed]
5. Guo, Y.; Li, C.; Zhao, H.; Gao, M.; Wang, Q. The Performance of Microfiltration Process for Purifying Lactic Acid in the Fermented Broth of Kitchen Waste. *Membranes* **2023**, *13*, 280. [CrossRef] [PubMed]
6. Guo, Y.; Li, C.; Zhao, H.; Wang, X.; Gao, M.; Sun, X.; Wang, Q. The Performance of Ultrafiltration Process to Further Refine Lactic Acid from the Pre-Microfiltered Broth of Kitchen Waste Fermentation. *Membranes* **2023**, *13*, 330. [CrossRef] [PubMed]
7. Kozmai, A.; Porozhnyy, M.; Gil, V.; Dammak, L. Phenylalanine Losses in Neutralization Dialysis: Modeling and Experiment. *Membranes* **2023**, *13*, 506. [CrossRef] [PubMed]
8. Cheng, Y.; Cheng, Q.; Zhao, C.; Ren, X.; Wang, Y.; Kou, Y.; Chon, K.; Ko, M.-H.; Hwang, M.-H. Evaluation of Efficiently Removing Secondary Effluent Organic Matters (EfOM) by Al-Based Coagulant for Wastewater Recycling: A Case Study with an Industrial-Scale Food-Processing Wastewater Treatment Plant. *Membranes* **2023**, *13*, 510. [CrossRef] [PubMed]
9. Kang, Y.-J.; Kim, Y.-J.; Yoon, S.-J.; Seo, D.-J.; Cho, H.-R.; Oh, K.; Yoon, S.-H.; Park, J.-I. Effective Removal of Acetaldehyde Using Piperazine/Nitric Acid Co-Impregnated Bead-Type Activated Carbon. *Membranes* **2023**, *13*, 595. [CrossRef]

membranes

Article

Recovery of Extracellular Polymeric Substances from Excess Sludge Using High-Flux Electrospun Nanofiber Membranes

Da-Qi Cao [1,*], Xiao-Dan Liu [1,†], Jia-Lin Han [1,†], Wen-Yu Zhang [2], Xiao-Di Hao [1], Eiji Iritani [3] and Nobuyuki Katagiri [4]

1 Sino-Dutch R&D Centre for Future Wastewater Treatment Technologies/Key Laboratory of Urban Stormwater System and Water Environment, Beijing University of Civil Engineering and Architecture, Beijing 100044, China
2 Institute of Soil Environment and Pollution Remediation, Beijing Municipal Research Institute of Environmental Protection, Beijing 100037, China
3 Department of Chemical Engineering, Nagoya University, Furo-cho, Chikusa-ku, Nagoya 464-8603, Japan
4 Department of Environmental Technology, Meijo University, 1-501 Shiogamaguchi, Tempaku-ku, Nagoya 468-8502, Japan
* Correspondence: caodaqi18@163.com or caodaqi@bucea.edu.cn; Tel.: +86-10-6832-2123
† These authors contributed equally to this work.

Abstract: The recycling of extracellular polymeric substances (EPSs) from excess sludge in wastewater treatment plants has received increasing attention in recent years. Although membrane separation has great potential for use in EPS concentration and recovery, conventional membranes tend to exhibit low water flux and high energy consumption. Herein, electrospun nanofiber membranes (ENMs) were fabricated using polyvinylidene fluoride (PVDF) and used for the recovery of EPSs extracted from the excess sludge using the cation exchange resin (CER) method. The fabricated ENM containing 14 wt.% PVDF showed excellent properties, with a high average water flux (376.8 L/(m^2·h)) and an excellent EPS recovery rate (94.1%) in the dead-end filtration of a 1.0 g/L EPS solution at 20 kPa. The ENMs displayed excellent mechanical strength, antifouling properties, and high reusability after five recycles. The filtration pressure had a negligible effect on the average EPS recovery rate and water flux. The novel dead-end filtration with an EPS filter cake on the ENM surface was effective in removing heavy-metal ions, with the removal rates of Pb^{2+}, Cu^{2+}, and Cr^{6+} being 89.5%, 73.5%, and 74.6%, respectively. These results indicate the potential of nanofiber membranes for use in effective concentration and recycling of EPSs via membrane separation.

Keywords: extracellular polymeric substance; recovery; electrospun nanofiber membrane; dead-end filtration; heavy-metal ion

Citation: Cao, D.-Q.; Liu, X.-D.; Han, J.-L.; Zhang, W.-Y.; Hao, X.-D.; Iritani, E.; Katagiri, N. Recovery of Extracellular Polymeric Substances from Excess Sludge Using High-Flux Electrospun Nanofiber Membranes. *Membranes* **2023**, *13*, 74. https://doi.org/10.3390/membranes13010074

Academic Editor: Chii-Dong Ho

Received: 17 December 2022
Revised: 29 December 2022
Accepted: 5 January 2023
Published: 7 January 2023

1. Introduction

Resource recovery from wastewater has received increasing attention for application in sustainable wastewater treatment systems [1,2]. In such systems, the handling of excess sludge accounts for almost half of the total investment and operational costs of wastewater treatment plants (WWTPs) [3–5], wherein excess sludge is the main source of resource materials [6–11]. If these resource materials can be recycled, the burden of sludge treatment will be reduced, thereby contributing to a circular economy and a circular society [7,10–12].

As a high-value-added product of excess sludge, extracellular polymeric substances (EPSs) are derived from the secretion of microbial cells and cell autolysis, and they mainly consist of polysaccharides, proteins, lipids, humus, nucleic acids, and DNA, accounting for 10–40% of the sludge dry weight [8,13]. The recovered eco-friendly EPSs can be used as adsorbents for heavy-metal ions [7,10,14], flame retardants [15,16], soil conditioners [17], and bioflocculants [18,19]. In particular, the EPSs in aerobic granular sludge (AGS), a recent innovation in biofilm reactors for wastewater treatment, are mainly composed of alginate;

they have higher added value than the commonly used EPSs recovered from various sludge in conventional wastewater treatments, such as the sequencing batch reactor technology, as well as the anaerobic or anoxic/oxic, and anaerobic/anoxic/oxic processes [1,6,9,11]. In addition, the removal of EPSs from excess sludge is conducive to sludge concentration and dehydration, agricultural use, incineration, and subsequent disposal [8,17,19]. However, the high moisture contents of EPS solutions extracted from sludge constitute the main bottleneck in the recovery process; thus, it is necessary to explore new technologies for EPS concentration [7,10].

Direct drying methods, such as heating and drying, are generally undesirable for this process due to their high energy consumption and potential generation of secondary pollution, not to mention the possibility that the polymer structures can be damaged during such processes [7]. Membrane separation technologies, therefore, represent possible alternatives for the separation and purification of such substances because of their high separation efficiencies and lack of secondary pollution [20–22]. In previous studies, the effective separation of polysaccharides and proteins was achieved by ultrafiltration [7,9], and the effective recovery of EPSs without damage or the generation of secondary pollution was possible [10]. However, EPSs are mixtures of various macromolecular substances; therefore, depending on the membrane characteristics, the EPSs can produce greater degrees of membrane fouling via pore clogging, foulant adhesion, foulant layer formation, and consolidation [23]. In addition, traditional filtration membranes possess a number of inherent limitations, such as a low flux and a susceptibility to pollution due to the regular void arrangement on the transverse membrane surface and the vertical pore arrangement [24,25].

To address such issues, the electrospinning approach has been considered for membrane preparation. This process involves the application of a strong electric field to generate nanofibers from a charged polymer solution [26–30]. As a result, electrospun nanofiber membranes (ENMs) with random interconnected pores can be formed from overlapping fibers [31–33], which provide interconnected structures with high porosities that allow liquids to pass through multiple channels [34,35]. In addition, ENMs tend to exhibit a superior fouling resistance to traditional membranes, thereby mitigating the blockage of a single membrane pore and resulting in more promising filtration applications [24,36,37]. For example, a polyvinylidene fluoride (PVDF) ENM with pore sizes <200 nm was used to fabricate the bases of thin-film composite membranes, giving a high flux of 993 L/(m^2·h), which is three times that of the conventional casting membrane [38]. Another PVDF nanofiber membrane fabricated by three-dimensional printing and near-field electrospinning showed an excellent filtration performance with a high flux of 1020.7 L/(m^2·h) and a particle rejection of 96.7% when an aqueous dispersion of SiO_2 (average particle size = 50 μm) was employed as the feed solution [32]. Even for natural river water, a high flux of 20,455 L/(m^2·h) was obtained at a rejection rate of 99.8% for river particles measuring ~2 μm in diameter using electrospun fibrous cellulose acetate membranes [39]. Furthermore, a nanoparticle rejection rate of ~99.0% was achieved in the filtration of nanoparticle suspensions [40]. Overall, ENMs have been shown to exhibit superior mechanical properties and high specific surface areas, in addition to permitting the facile modification of their functional groups to provide unique functionalities [34,41], thereby rendering them excellent filtration materials [42].

The unique nanostructures of ENMs not only facilitate membrane modification but also provide the conditions for the formation of free nanogaps with directional nanochannels to permit substance transport, thereby resulting in the selective separation and concentration of polymeric substances [22,42–44]. For example, an ENM fabricated by the hot-pressing of PVDF and polyacrylonitrile presented high rejection percentages for bovine serum albumin and bovine γ-globulin molecules, i.e., 60% and 75%, respectively [43]. In addition, a nanofibrous membrane composed of a hollow porous polystyrene/ethylene–vinyl acetate copolymer blend was generated by electrospinning and chemical modification and was subsequently used to separate and purify proteins with a 94.35% rejection rate of serum albumin [22]. Furthermore, the excellent separation performances of ENMs are of particular interest in the medical field [42]. For example, a dual-layer nanofiber membrane composed of a polydopamine/polyacrylonitrile layer and a chitosan/sericin layer achieved creatinine and urea clearance rates of 82.3 and 92.8%, respectively, while retaining 83.9% of bovine serum albumin [37]. Moreover, a polymeric hemodialysis membrane composed of an electrospun polyacrylonitrile nanofibrous support layer achieved a 98.4% retention of bovine serum albumin after 4 h [45]. In addition, PVDF, which has excellent mechanical and thermal strength, is a semi-crystalline polymer mainly composed of 59.4% fluorinated chains and synthesized by free-radical polymerization [31]. Furthermore, these fluorinated chains on the surface of PVDF impart chemical and aging resistance [30,46]. Due to these characteristics, PVDF has been widely used in the development of novel nanofiber membranes [27,30–32,35,38,41,43,44]. Table 1 lists the typical parameters associated with a selection of ENMs fabricated using PVDF over the past 5 years.

Table 1. Typical parameters for previously reported ENMs fabricated using PVDF over the last 5 years and the PVDF ENMs fabricated in this study *. *d*: pore size; *D*: fiber diameter; ω: porosity; *T*: thickness; θ: water contact angle; F_t: tensile strength; ε: nominal tensile strain at break.

Membrane Fabrication			*d*	*D*		*T*		F_t		
PVDF [wt.%]	**Solvent [*v*/*v*]**	**Modification [wt.%]**	**[μm]**	**[μm]**	**ω [%]**	**[μm]**	**θ [°]**	**[MPa]**	**ε [%]**	**Ref.**
15	DMF/acetone: 9:1	None	1.82	0.128	68	103	90	1.86	55.2	[47]
		w/*h*: 1:1	30		36		137	0.19	4.8	
15	DMAc/acetone: 4:1	2:1	45	-	40	-	134	0.13	3.4	[32]
		3:1	90		44		131	0.08	2.4	
		GO: 0	0.8	0.167			140	6.40	49.3	
15	DMF/acetone: 3:1	0.5	0.99	0.259	-	100	131	5.20	27	[48]
		1	1.13	0.311			119	4.80	12.6	
		3	1.21	0.355			110	4.40	9.5	
15	DMF	SiO_2NPs: 0	22.00	0.400	-	-	<135	-	-	[49]
		4	22.00	0.600			135			
		None	1.33		84	121	92	2.24	13.0	
15	DMAc	f-SiO_2NPs: 1	1.24	-	80	116	152	2.13	26.0	[50]
		AgNPs/f-MWCNTs: 2:1	0.64		82	135	56	2.38	30.0	
		SiO_2NPs: 0	0.37	0.265	84	107	132	1.83		
13.5	DMF/acetone: 6:4	2	0.32	0.206	81	110	143	2.16	-	[41]
		4	0.28	0.209	78	97	149	2.56		
		6	0.27	0.209	77	98	155	1.55		
	DMF/acetone: 6:4		1.30	0.415		91	129			
12	DMSO/acetone: 6:4	None	2.40	0.625	-	87	130	-	-	[27]
	Anhydrous/acetone: 6:4		3.58	0.521		82	116			
12		PAN: 0		0.550			133	7.125	13.5	
10.8	DMF	1.2	-	0.600	-	-	131	9.95	13.0	[51]
6		6		0.550			118	7.40	7.8	
		DTPA/MWCNT/TiO_2: 0	0.006	0.120		114	62	0.13	1.8	
8	NMP	0.5	-	0.116	-	117	47	0.26	3.5	[31]
		2		0.102		118	38	0.29	5.7	
		6	0.007	0.096		118	27	0.21	6.1	
8	DMF/acetone: 6:4	GO: 0	0.3	0.176	83–85	50	137	-	-	[52]
		5	0.19	0.131			89			
14			1.47	0.504	52.46	125	109.5	1.54	15.30	
18	DMAc/acetone: 4:1	None	7.66	0.865	71.53	186	119.1	2.22	20.16	This study
22			19.36	1.772	73.24	240	129.3	2.58	36.78	

* DMF: dimethylformamide; NMP: *N*-methylpyrrolidone; DMSO: dimethyl sulfoxide; DMAc: dimethylacetamide; NPs: nanoparticles; DTPA/MWCNT/TiO_2: diethylenetriaminepentaacetic acid (DTPA)-functionalized multi-walled carbon nanotube (MWCNT)/TiO_2; *w*/*h*: length–width ratio for 3D printing pore geometry; GO: graphene oxide; f-SiO_2NPs: silanized silica nanoparticles; AgNPs/f-MWCNTs: silver nanoparticles and carboxylated multiwalled carbon nanotubes; PAN: polyacrylonitrile.

Heavy metals are responsible for reducing the quality of water owing to their high toxicities, their ability to bioaccumulate, and their nonbiodegradable nature [10,53,54]. Thus, ENMs have been examined for their application in the removal of heavy metals [55,56]. For example, metal–organic frameworks were enmeshed in PVDF and polyacrylonitrile electrospun nanofibers to produce nanofibrous membranes for the treatment of Pb^{2+} solutions [57], while PVDF nanofibrous membranes hybridized with silicon oxide nanoparticles exhibited a Cu^{2+} adsorption of ~21.9 mg/g [58]. In addition, the incorporation of carbohydrate polymers, such as chitosan, cellulose, alginate, pullulan, starch, and hyaluronic acid into nanofibrous structures via the electrospinning approach has also been proven effective in the adsorption of heavy metals [49,59]. More specifically, a polydopamine-containing PVDF nanofiber membrane achieved a Cu^{2+} adsorption capacity of 26.7 mg/g [60], in addition to an adsorption capacity of 126.7 mg/g for Cr^{6+} [61]. Furthermore, a multilayer electrospun nanofibrous membrane composed of chitosan and polyvinyl alcohol was used for the dynamic removal of Cu^{2+} from synthetic wastewater, giving a rejection rate of up to 98.6% [62].

Herein, the novelty of the present study is to first apply PVDF ENMs with a high water flux, to separate and concentrate EPSs that are extracted from the excess sludge using the cation exchange resin (CER) method. The simplest filtration process, i.e., dead-end filtration (DEF), was carried out to eliminate influence of hydraulic conditions. Initially, the key properties of the PVDF ENMs are analyzed; subsequently, the effects of different PVDF mass fractions on the EPS recovery efficiency and the water flux are investigated and compared with those of typical commercial filtration membranes. In addition, the influence of the applied pressure on the ENM filtration of EPS is also investigated, along with the fouling and reusability of the ENMs. Lastly, the removal of heavy-metal ions (HMIs) via DEF with the EPS filter cake formed on the ENM surface (EPS–ENM–DEF) is confirmed and evaluated, and the advantages of removing HIMs using EPS–ENM–DEF are analyzed.

2. Materials and Methods

2.1. Materials

Polyvinylidene fluoride (PVDF, HSV900, $M_W \approx 1000$ kDa, Arkema Inc., Paris, France) was used as the electrospinning polymer. *N,N*-Dimethylacetamide (DMAc, analytical reagent, 99.5%) and acetone (analytical reagent, 99%) were used as organic solvents and were purchased from the Rhawn Chemical Reagent Co., Ltd., and the Beijing Reagent Co., Ltd., respectively. $PbCl_2$, $CuCl_2{\cdot}H_2O$, and K_2CrO_7 (analytical grade) were purchased from Sigma Aldrich (St. Louis, MO, USA). H_2SO_4 (98 wt.%), HNO_3 (98 wt.%), and phenol (C_6H_5OH) were obtained from the Beijing Reagent Co., Ltd. Ultrapure water (resistivity $\geq$18.2 MΩ) was obtained by purifying tap water using an Arium Comfort II ultrapure water system for laboratory use (Sartorius Corp., Göttingen, Germany). The cellulose ultrafiltration membranes (molecular weight cutoffs [MWCOs] = 10 and 100 kDa) were purchased from Millipore Corp. (Billerica, MA, USA). Filter paper (pore diameters = 4 and 7 μm) was purchased from Advantec Corp., Tokyo, Japan. The disposable polyethersulfone filter membrane (0.45 μm) was purchased from the Jinteng Experimental Equipment Co., Ltd., Tianjin, China. The cation exchange resin (CER, Amberlite IR 120 Na) was obtained from Rohm and Haas Corp., Philadelphia, PA, USA. The dialysis bag (MWCO = 3500) was obtained from Viskase Corp., Lombard, IL, USA. Excess sludge was obtained from Beijing Dongba WWTP.

2.2. Fabrication of the PVDF Nanofiber Membranes

The PVDF nanofiber membranes were prepared via the electrospinning technique (ET2535, Beijing Yongkang Leye Co., Ltd., Beijing, China) with a high-voltage supply of 10–12 kV. The PVDF powder (1.5 g; 14, 18, and 22 wt.%) was initially dissolved in a mixture of DMAc and acetone with a volume ratio of 4:1 (the volumes of the mixture were 8.0, 6.0, and 4.6 mL, respectively) at 300 rpm and 60 °C for 4 h in a constant-temperature water bath (Changzhou Gaode Instrument Manufacturing Co., Ltd., Changzhou, China). Subsequently, each solution was defoamed using a circulating water vacuum pump (SHZ-D(III), GongYi city RuiDe Instrument and Equipment Co., Ltd., Zhengzhou, China) and cooled prior to electrospinning. The desired polymer solution was then transferred into a plastic syringe and delivered at a rate of 4 mL/h using a syringe pump (G20, ID = 0.60 mm). The applied voltage was optimized and fixed at 10.5 kV (+8.0/−2.5 kV) for each membrane fabrication. The drum was covered with aluminum foil at a distance of 20 cm from the syringe needle, and the temperature and humidity conditions inside the electrospinning chamber were set at 22–25 °C and 38–42% relative humidity, respectively. The samples were then cleaned repeatedly with deionized water to remove any residual solvent and give the desired PVDF nanofiber membranes. Finally, the prepared sample was cut into a circular film with a diameter of 6 cm using a custom circular cutter. The fabrication process is illustrated in Figure 1a. The obtained ENMs were denoted as ENM-14, ENM-18, and ENM-22 according to the wt.% of PVDF employed for their preparation. It should be noted that the pre-experiments were conducted under various conditions, such as a PVDF mass ratio of 10–24 wt.% and a spinning rate of 4–10 mL/h; however, only the key results are presented in the manuscript.

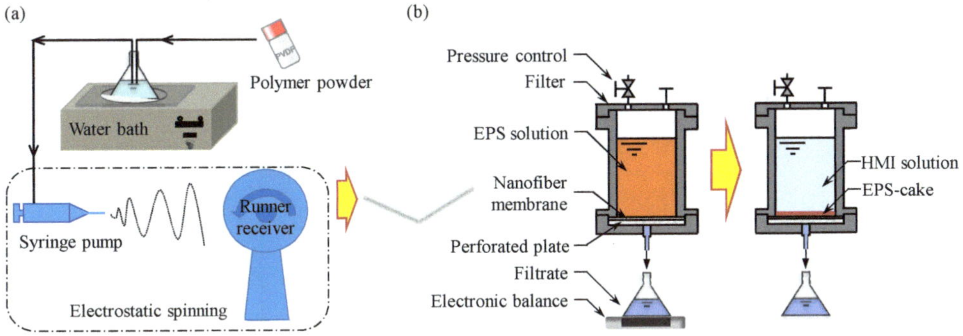

Figure 1. Schematic representations of (**a**) the nanofiber membrane fabrication process, and (**b**) the dead-end filtration apparatus and the experimental process employed herein.

2.3. *Sample Preparation*

The EPSs were extracted from the excess sludge sample via CER method [7]. Excess sludges were centrifuged at 4000× g and 4 °C for 20 min. Subsequently, the supernatant was removed, and the sediment was collected for experimental analysis and freeze-dried. First, sludge suspensions were prepared by dispersing 1 g of the obtained dry sludge samples in 250 mL of ultrapure water. The corresponding resin (70 g/g volatile solids; VS) was placed into the sludge suspension, and the mixture was stirred at 500 rpm for 4 h. Subsequently, the sludge suspensions were centrifuged at 4000× g and 4 °C for 20 min. Each supernatant was dialyzed in a dialysis bag (MWCO = 3500 Da) against a ninefold volume of DI water for 24 h, and the process was repeated two more times to remove all impurities such as various heavy-metal ions that could have been present. The sediment was discarded. Finally, EPS powder was produced by freeze-drying the solution in the dialysis bag. Notably, the EPS powders produced were afforded in quantities that were sufficient for the experiments performed, allowing the prepared batches to be used in all experiments, thus ensuring consistent EPS characteristics.

It was confirmed that this technique allowed recovery of the EPSs because of negligible destruction to the microbial cells in the excess sludge [7,10]. Therefore, these samples can be considered as mixed water-soluble polymers. The EPS was dissolved in ultrapure deionized (DI) water at 24 °C and stirred for 4–12 h to obtain the desired EPS solution. All samples were used immediately to prevent any errors caused by microbial consumption of the organic matter present in the aqueous solution [23,63].

2.4. Two-Stage ENM Filtration

The EPS solutions were filtered using a custom-made pressure filter device, as shown in Figure 1b. The two-stage filtration process was carried out as described previously [10]. Herein, the first step involved concentration of the EPSs, wherein the fabricated ENMs were used to filter the EPS solutions (50 mL) to form a dense EPS filter cake layer on the surface of the membrane at a filtration pressure (p_1) of 20 kPa. An ENM filter membrane with an EPS cake layer was obtained as a result of this process. The second step involved the removal of HMIs. More specifically, heavy-metal-containing wastewater samples (10 μM solutions of Pb^{2+}, Cu^{2+}, and Cr^{6+}) were filtered through the cake at a filtration pressure (p_2) of 20 kPa. The HMIs were removed by the synergistic effect of the EPS filter cake and the ENM. For comparison, the ENMs were also used to remove the HMIs without the presence of an EPS filter cake, which can be considered to reflect the adsorption of HIMs by the ENM. Notably, to eliminate the effects of acid, base, or buffers, the pH was not adjusted in any experiments; it was maintained at 6.5 obtained using a pH meter (Mettler Toledo FE20).

2.5. Analytical Methods

2.5.1. Membrane Characterization

The morphological properties of the membrane surface and the cross-sectional areas of the PVDF nanofiber membranes were characterized using scanning electron microscopy (SEM, G300, Zeiss, Jena, German) at 2–3 kV. All samples were cut into dimensions of 1 cm × 1 cm and were pretreated with gold spraying for 10 min under vacuum. The electron microscope images were observed at 5–50 k. Image Pro-Plus 6.0 software was used to visualize the SEM images and give the fiber diameters of the ENMs. The porosities and pore sizes of the membranes were obtained by mercury porosity experiments (Auto-Pore Iv 9510, Micromeritics Instrument Co., Ltd., Norcross, GA, USA), while a thickness gauge (HCC–18, Liuling Instrument Factory, Shanghai, China) was used to measure each membrane thickness. The water contact angles of the PVDF ENMs were measured using a contact angle meter (JC2000D4M, Zhongchen Digital Technology Equipment Co., Ltd., Shanghai, China); finally, the mechanical properties of each sample (shaped into strips of dimensions 2 cm × 1 cm) were tested using an electronic universal testing machine (nonmetallic direction) (Inspekt table blue 5 kN, Hegewald & Peschke, Nossen, Germany) at a distance of 20 cm and a tensile rate of 2 mm/min.

2.5.2. Evaluation of the Membrane Resistance and Filtration Behaviors

All experiments employed DEF with an effective area of 19.6 cm^2 using a self-made pressure filter device (Figure 1b). A constant pressure was applied using N_2 gas and was controlled using an automatic pressure-regulating valve. The filtrate liquor was collected in a reservoir and placed on an electronic balance connected to a personal computer to record the mass versus time data. The weights were converted to volumes using a density correlation (i.e., mass/density = volume). The filtration rate was obtained through a numerical differentiation of the volume versus time data. The resistance properties of the membranes were examined by means of ultrapure water filtration experiments [6]. The membrane resistance, R_m was calculated using Equation (1) as follows:

$$R_{\mathrm{m}} = \frac{p}{\mu \cdot J}, \tag{1}$$

where p represents the applied filtration pressure, μ is the viscosity of water, and J is the filtration rate during the ultrapure water membrane filtration process.

During cake filtration, the relationship between the reciprocal of the filtration rate ($d\theta/dv$) and the cumulative filtrate volume collected per unit effective membrane area, v, is linear in accordance with the Ruth filtration rate equation [6,7]:

$$\frac{d\theta}{dv} = \frac{2}{K_v} v + \left(\frac{d\theta}{dv}\right)_m, \tag{2}$$

where θ is the filtration time, and K_v is the Ruth filtration coefficient, indicating the filterability of the feed. In addition, $(d\theta/dv)_m$ represents the membrane flow resistance and is an intercept on the $(d\theta/dv)$ axis according to the Ruth plots of $(d\theta/dv)$ as a function of v, which is the value of the reciprocal of the filtration rate at the start of filtration prior to cake formation. In this study, the behaviors of the flux decline were analyzed by evaluation of the curved shapes of the Ruth plots ($d\theta/dv$ vs. v) [64–66].

2.5.3. Reusability of the Nanofiber Membranes

To evaluate the reusability of each nanofiber membrane, equivalent-concentration EPS solutions were filtered through each PVDF nanofiber membrane under a constant pressure of 20 kPa for a total of five times. Between each filtration, the EPS filter cake was removed by physical scraping. The polluted membrane was then placed in a sealed bag. and ultrasonic cleaning (600 W) was carried out for 10 min at 25 °C. It was observed that the membrane fouling was almost completely removed probably because the recovered EPSs obtained via CER were water-soluble. Prior to each subsequent filtration, pure water was passed through the membrane to check for channel blockage, and it was found that R_m remained relatively constant.

2.5.4. Size Distribution, Fourier-Transform Infrared Spectroscopy, and Zeta Potential

The typical size distribution of the colloidal EPS solution was measured by dynamic light scattering using a Zetasizer Nano ZS90 size analyzer (Zetasizer Nano ZS90, Malvern Co., Ltd., Malvern, UK). To prepare the samples for analysis by Fourier-transform infrared (FTIR) spectroscopy, the EPS sample was obtained by vacuum freeze-drying (FD-1A-50, Beijing Boyikang Laboratory Instruments Co., Ltd., Beijing, China). The samples were then mixed with KBr (1:100 mass ratio) and dried overnight at 120 °C. Finally, the dried samples were ground into a powder using an agate mortar and pestle, and a tablet press was used to prepare the KBr discs required for functional group analysis by FTIR spectroscopy (Nicolet is5, Thermo Scientific, Waltham, MA, USA). The zeta potentials of the EPSs and the membrane surfaces were measured using a solid surface zeta potential analyzer (SurPASS^{TM} 3 Eco, Anton Paar, Graz, Austria). The surface zeta potential was determined as a function of pH in a 0.001 M KCl electrolyte solution upon varying the solution pH from 5 to 7 by the addition of a HCl or NaOH solution through the automatic titration unit of the instrument [67].

2.5.5. Determination of the EPS Recovery and the HMI Removal Efficiencies

A 5 mL portion of the collected filtrate was used to determine the concentrations of the effluent EPSs and HMIs present in the filtrates obtained by ENM filtration and EPS–ENM filtration, respectively. The concentration of polysaccharides (C_p) in the EPS solution or the collected filtrate obtained using different membranes was determined according to the phenol/sulfuric acid method using UV/Vis spectrophotometry (Cary 5000, Agilent Technologies Co., Ltd., Waldbronn, Germany) (UV method) [8]. As a control, the EPS concentration in the EPS solution or the collected filtrate (C_e) was obtained using the standard freeze-drying weighing method [7]. The HMI concentrations (C_i) in the aqueous solutions were measured using inductively coupled plasma (ICP) spectrometry (ICP 7000 Series, Thermo Scientific, Waltham, MA, USA) after mixing the samples with HNO_3 (1% v/v) and filtering through a 0.45 μm membrane to meet the sampling requirements. All experiments were repeated at least twice. Thus, the polysaccharide recovery efficiency ($\eta_p = 1 - C_p/C_{p0}$), EPS recovery efficiency ($\eta_e = 1 - C_e/C_{e0}$), and HMI removal efficiency ($\eta_i = 1 - C_i/C_{i0}$) were obtained. In these equations, C_{p0} is the initial polysaccharide concentration in the initial EPS solution, C_{e0} is the initial EPS concentration, and C_{i0} is the initial concentration of the HMI in the wastewater sample.

The average EPS recovery rates (η_{av}) during DEF of the EPS solution using the fabricated ENMs and commercial membranes were evaluated using both the UV and the weighing methods, wherein the evaluated data were obtained for the cumulative filtrate volume per unit membrane area, v = 2 cm. In Figure S1, the average EPS recovery rates obtained using the UV method are compared with those obtained using the weighing method. As shown in the figure, all data fell within the range bounded by the error bars of ±10% depicted by the dotted lines, indicating that the UV method could accurately evaluate the average EPS recovery rate in the DEF of the EPS solution. Hence, the values of η_{av} were evaluated using the UV method for the purpose of this study.

3. Results and Discussion

3.1. Characterization of the Nanofiber Membranes

Table 1 also lists the typical morphological and physical parameters of the PVDF ENMs fabricated in this study. In addition, Figure 2 shows the surface morphologies of the ENMs, indicating that a larger PVDF mass fraction resulted in a larger fiber diameter and membrane pore size. More specifically, the fiber diameters and pore diameters of the ENMs were obtained by combining SEM with Image Pro 6.0 image processing software, wherein, as shown in Table 1, the porosity (52.46–73.24%) and the fiber diameter (0.504−1.772 μm) were confirmed to increase with an increase in the PVDF mass fraction [68]. Since the hydrophobic properties of a membrane are known to affect the filtration resistance [69], the water contact angles (WCAs) of the ENMs were measured and are listed in Table 1 (see also Figure S2). As indicated, the WCA increased from 109.5° to 129.3° upon increasing the PVDF mass fraction, and the average WCA was 119.3°; this value corresponds to a typical hydrophobic membrane and results from the high concentration of PVDF fluorine atoms on the ENM surface [70]. Notably, except for the properties of membrane material, the membrane surface roughness also affected the WCA; herein, the porosity of membranes increased with increasing the PVDF mass fraction, resulting in numerous air pockets in the membrane surfaces, leading to higher hydrophobicity (larger WCA) [41].The tensile stress–strain curves of the fabricated ENMs are shown in Figure S3 and Table 1, wherein it can be seen that the tensile strength and nominal tensile strain at break of the ENMs increased upon increasing the PVDF mass fraction because of large fiber diameter and membrane thickness. Of the various samples prepared, ENM-22 exhibited the best mechanical properties due to the higher loading of PVDF, which increased the tensile stress that could be supported by the membrane.

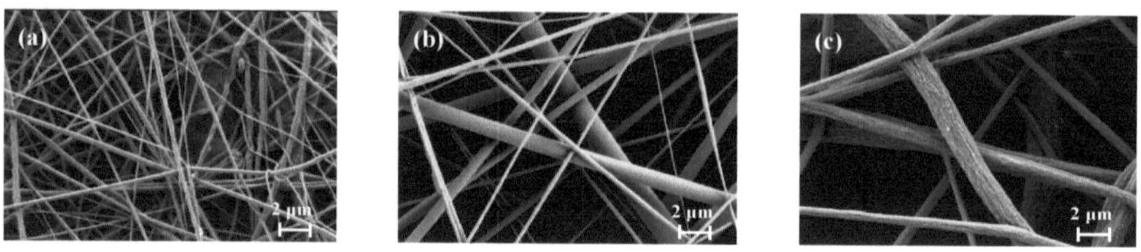

Figure 2. Typical SEM images of the nanofiber membranes obtained using PVDF mass fractions of (**a**) 14 wt.%, (**b**) 18 wt.%, and (**c**) 22 wt.%, denoted as ENM-14, ENM-18, and ENM-22, respectively.

3.2. Comparison between the EPS Recovery Properties of the Fabricated Nanofiber Membranes and Commercial Membranes

The average water flux (J_{av}) and average EPS recovery rate (η_{av}) obtained during the DEF of the EPS solution using the fabricated ENMs and commercial membranes are shown in Figure 3, wherein the evaluated data were obtained for the cumulative filtrate volume per unit membrane area, v = 2 cm. In general, the nanofiber membranes are ultrafiltration or microfiltration membranes; therefore, herein, the most widely used ultrafiltration membranes (MWCO = 10 and 100 kDa, hydrophobic) and microfiltration membranes (4 and 7 μm, hydrophilic) from Millipore Corp. and Advantec Corp., respectively, were compared. In the case of ENM-14, the highest value of J_{av} was achieved (i.e., 376.8 L/(m^2·h)) in addition to a high η_{av} value of 94.1%, while, for the 10 kDa commercial membrane that was previously reported to exhibit a high EPS recovery efficiency (>90%) [7], a significantly lower J_{av} of 10.7 L/(m^2·h) was recorded, although the η_{av} value was slightly higher (i.e., 96.4%). These results clearly indicate that the water flux exhibited by ENM-14 was ~35.2 times higher than that of the 10 kDa commercial membrane. It is worth noting that our ENMs were prepared using low-cost PVDF materials, unlike in the cases of commercial ultrafiltration and microfiltration membranes. Therefore, in addition to their high porosities and highly interconnected pore structures that offer a low hydraulic resistance for water transportation and yield a high throughput with a low energy consumption, the ENMs reported herein also impart economic benefits during the recovery and concentration of EPSs and other substances.

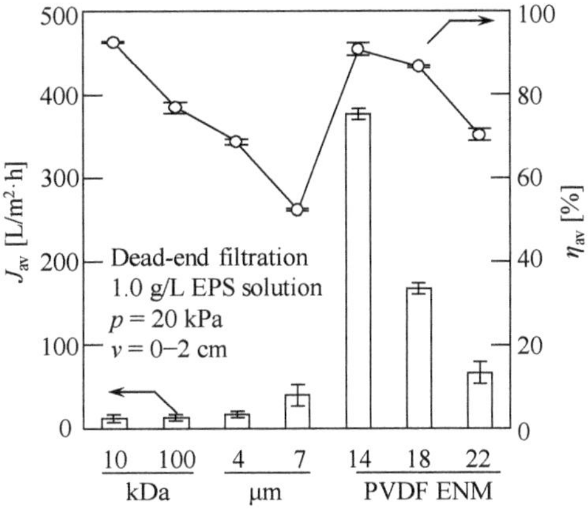

Figure 3. Average water flux (J_{av}) and average EPS recovery rate (η_{av}) in the dead-end filtration of a 1.0 g/L EPS solution. A filtration pressure of 20 kPa was used, and the evaluated data were obtained for the cumulative filtrate volume per unit membrane area, v = 2 cm. Millipore ultrafiltration membranes (MWCO = 10 and 100 kDa), Advantec filter paper (4 and 7 μm), and the prepared PVDF ENMs (PVDF mass fractions of 14, 18, and 22 wt.%) were evaluated.

Figure 4 shows the resistance values of the ENMs and the commercial membranes. As indicated, for the four commercial membranes, R_m decreased with increasing membrane pore size, which accounted for the increasing and decreasing J_{av} and η_{av} values, respectively, as shown in Figure 3. In contrast, for the nanofiber membranes, increases in both J_{av} and η_{av} values were observed upon decreasing the membrane resistance (see Figures 3 and 4). More specifically, the ENM-14 membrane exhibited both the highest water flux and the highest EPS recovery rate of the various samples investigated, and this may be due to the quasi-three-dimensional network structure of the cross-configuration [68]. Since the EPSs are considered a mixture of various polymers extracted from the excess sludge, large numbers of polysaccharides, proteins, DNA, and other substances are present [7,10]. As a result, the EPSs possess a wide particle size distribution of 6.2–226.7 μm, as shown in Figure S4. Thus, the smaller colloids easily enter the ENM pores, resulting in membrane blockage [65,66]. However, the pore size of ENM-14 with low porosity and small thickness appeared to be sufficiently small to retain the majority of these colloidal particles in the EPSs; in contrast, pore blocking occurred in the ENM-18 and ENM-22 membranes because both Ruth plots ($d\theta/dv$ vs. v) exhibited more remarkable downward convex curves (see Figure 5a) based on the membrane pore clogging model [65,66], thereby lowering their EPS recovery rates and water fluxes.

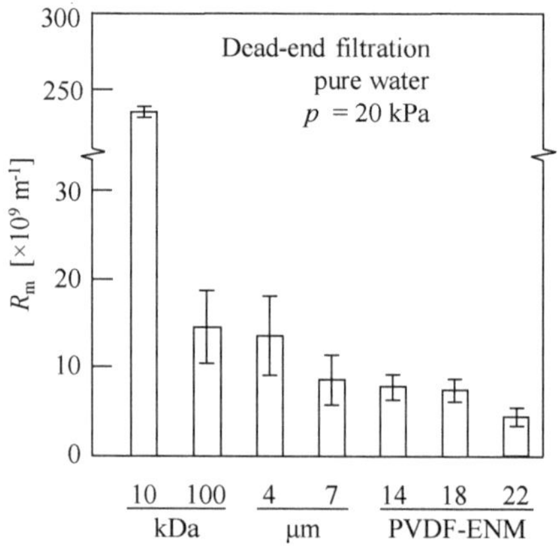

Figure 4. Membrane resistance (R_m) obtained via the dead-end filtration of ultrapure water at a filtration pressure (p) of 20 kPa. Millipore ultrafiltration membranes (MWCO = 10 and 100 kDa), Advantec filter paper (4 and 7 μm), and the prepared PVDF ENMs (PVDF mass fractions of 14, 18, and 22 wt.%) were evaluated.

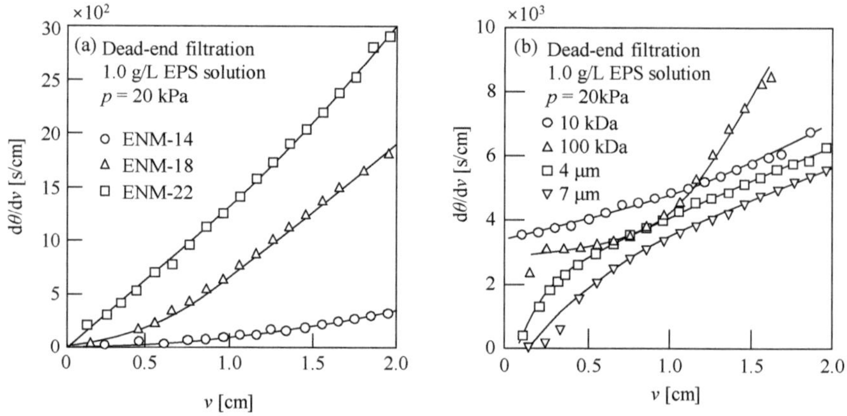

Figure 5. Filtration behaviors during the dead-end filtration of a 1.0 g/L EPS solution at a filtration pressure (p) of 20 kPa for (**a**) the ENMs, and (**b**) the commercial membranes. The horizontal axis (v) represents the cumulative filtrate volume per unit membrane area, the vertical axis ($d\theta/dv$) represents the reciprocal of the filtration rate, and θ is the filtration time.

Figure 5 shows the corresponding filtration behaviors of the ENMs and the commercial membranes, wherein it can be seen that the filtration rates of ENMs declined more slowly than those of the commercial membranes. More specifically, the obtained Ruth filtration plots [64], i.e., the relationship between the reciprocal of the filtration rate ($d\theta/dv$) and the cumulative filtrate volume collected per unit effective membrane area (v), show different shapes, i.e., downward convex curves for the ENMs (ENM-14, ENM-18, and ENM-22) and the commercial ultrafiltration membranes (10 and 100 kDa), and upward convex curves for the commercial microfiltration membranes (4 and 7 μm), indicating that the pore blocking mechanisms of nanofiber membranes differ from those of conventional commercial membranes.

3.3. Influence of the Applied Pressure on the ENM Filtration of an EPS Solution

In general, the filtration rate of a membrane can be enhanced by increasing the applied filtration pressure [7]; therefore, we moved on to examine the influence of the filtration pressure on the concentration and recovery of EPSs using the fabricated ENMs. Thus, Figure 6a shows the influence of p on J_{av} and η_{av} when the EPS solution was subjected to filtration through the ENM-14 specimen. As shown, the average water flux fluctuated upon increasing the pressure from 20 to 80 kPa in steps of 20 kPa. Overall, an increased water flux was observed; however, upon increasing the pressure further to 100 kPa, an overall decrease was observed. These observations were attributed to the high compressibility of the EPS filter cake [7] and blockage of the ENM pores at higher pressures [66]. However, it was also found that the average EPS recovery rate was not affected by the filtration pressure and remained at ~94%. In addition, Figure 6b shows the filtration behavior of a 1.0 g/L EPS solution at various filtration pressures (20–100 kPa), wherein the results confirm the lack of a proportional relationship between the filtration pressure and the water flux, and the Ruth filtration plots show downward convex curves regardless of the filtration pressure employed, which further reveals the membrane pore blocking based on the blocking filtration model [66]. Therefore, to reduce energy consumption, a low filtration pressure was proposed to concentrate and recycle EPSs using these fibrous membranes.

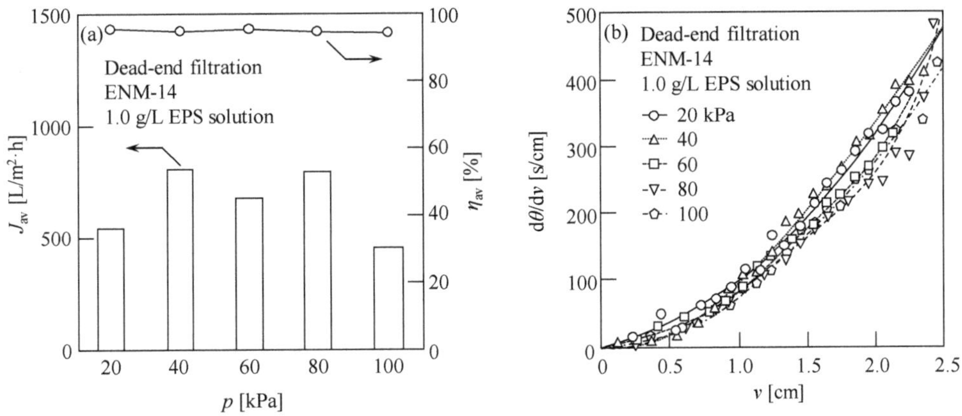

Figure 6. Using the ENM-14 specimen: (**a**) Average water fluxes (J_{av}), average EPS recovery rates (η_{av}), and (**b**) filtration behaviors in the dead-end filtration of a 1.0 g/L EPS solution at various filtration pressures (p = 20, 40, 60, 80, and 100 kPa). The evaluated data were obtained for the cumulative filtrate volume per unit membrane area, v = 2 cm. The horizontal axis shows v, while the vertical axis shows the reciprocal of the filtration rate ($d\theta/dv$), and θ is the filtration time.

3.4. Fouling and Reusability of the ENMs in the Filtration of EPSs

The prevention of membrane fouling is a key factor in the development of new membrane technologies. It is worth emphasizing that the EPSs used in this study were extracted from the excess sludge sample via CER, which is able to efficiently recover the EPSs because of little destruction of microbial cells in the excess sludge [7,10]. Therefore, the EPSs can be considered as mixed water-soluble polymers, and it was found that an EPS filter cake was formed on the membrane surface after filtration. The EPS filter cake was removed by physical scraping before the ultrasonic cleaning of the polluted membrane. Typical images of the ENMs before and after removal of the EPSs by ultrasonic cleaning (600 W, 10 min) are shown in Figure 7, and the polluted membrane with EPS filter cake can be clearly observed (see Figure 7a), indicating that the EPS fouling within the membrane pores and on the membrane surface can be removed using this simple technique because of their water solubility. Meanwhile, Figure 8 shows the zeta potentials of the EPSs and the membrane surfaces of the ENMs at pH values of 5, 6.5, and 7. As shown in the figure, both species exhibit a negative charge regardless of the pH, and the value of zeta potential increases with increasing pH, revealing that electrostatic repulsion between the EPSs and the membranes represents an alternative EPS filtration mechanism. Thus, the EPS cake layer formed in our study was completely removed via simple physical cleaning owing to the water solubility of EPSs and the electrostatic repulsion between the EPSs and the membranes.

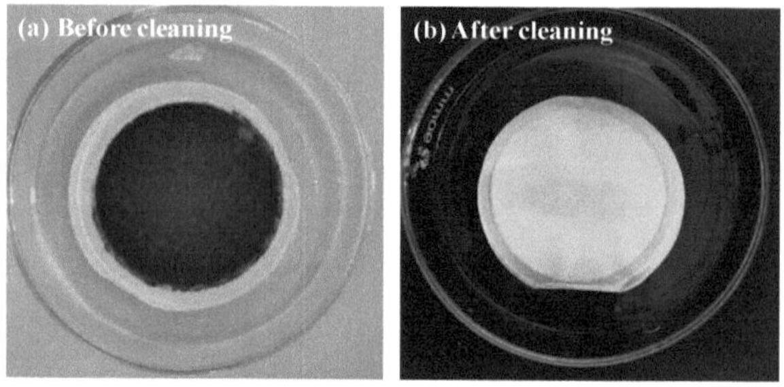

Figure 7. Typical photographic images of the ENMs with recycled and concentrated EPS (**a**) before and (**b**) after ultrasonic cleaning.

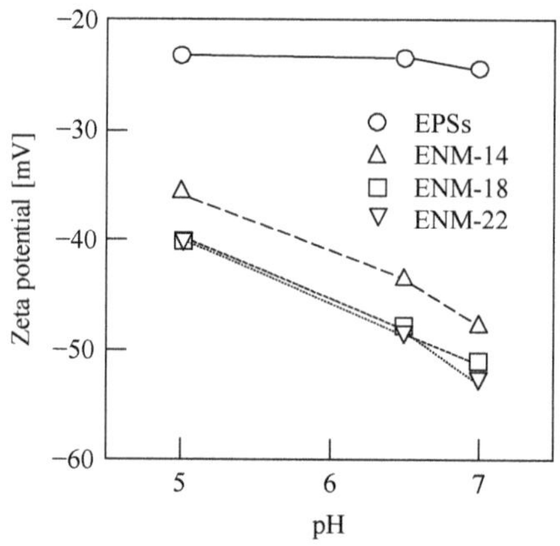

Figure 8. Potentials of the EPS and the ENM membrane surfaces at pH values of 5, 6.5, and 7.

Using the ultrasound-based membrane cleaning technique (i.e., ultrasonication for 10 min after each filtration), the filtration experiments were repeated five times using the recycled membrane to evaluate the reusability of the fabricated ENM. In this case, the ENM-14 and ENM-22 membranes were used because they exhibited the lowest and highest degrees of membrane fouling (i.e., the maximum and minimum water fluxes), respectively. Figure 9 shows the average water flux and average EPS recovery rate during the DEF of a 1.0 g/L EPS solution with a filtration pressure of 20 kPa using the ENM-14 and ENM-22 specimens. From this figure, it can be seen that, over five filtration cycles, the average EPS recovery remained relatively constant, as did the average water flux, although there was a slight increase in the latter. These results, therefore, indicate that the fabricated ENMs exhibit a good reusability for the recovery of EPSs extracted from excess sludge via CER.

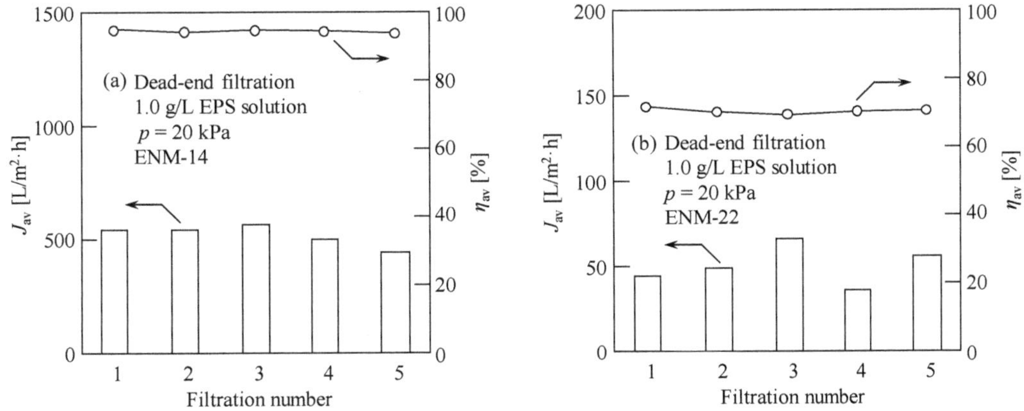

Figure 9. Average water flux (J_{av}) and average EPS recovery rate (η_{av}) during the dead-end filtration of a 1.0 g/L EPS solution with a filtration pressure (p) of 20 kPa using the (**a**) ENM-14 and (**b**) ENM-22 specimens. Five filtration cycles were carried out wherein the membrane was cleaned using ultrasonic irradiation for 10 min between filtrations. The evaluated data were obtained for the cumulative filtrate volume per unit membrane area, v = 2 cm.

3.5. Removal of HIMs via EPS–ENM Dead-End Filtration

Subsequently, we carried out a two-stage DEF procedure (see Figure 1b) using the prepared ENM-14 specimen, wherein the first stage involved the above-described recycling and concentration of an EPS solution, and the second stage involved the removal of HIMs from wastewater, i.e., the removal stage of HMIs via EPS–ENM–DEF. The detailed working mechanism for the removal of HIMs via EPS–ENM–DEF can be seen in our previous study [10]. Thus, the removal efficiency achieved using a 10 μM aqueous solution of the desired HIM (Pb^{2+}, Cu^{2+}, or Cr^{6+}, pH 6.2–6.7) by DEF in the presence of the EPS filter cake on the ENM surface (i.e., EPS–ENM–DEF) is shown in Figure 10. It should be noted that the EPS filter cakes were formed in the first stage from 50 mL aliquots of 0, 0.1, 0.2, and 1.0 g/L EPS solutions to give filter cake masses of 0, 5, 10, and 50 mg, respectively. As shown in the figure, the HMI removal efficiencies increased upon increasing the EPS filter cake mass, with removal efficiencies of 89.5%, 73.5%, and 74.6% being achieved for Pb^{2+}, Cu^{2+}, and Cr^{6+}, respectively, using the 50 mg EPS filter cake. Furthermore, adsorption experiments revealed membrane adsorption capacities of 20.6%, 21.8%, and 12.0% for Pb^{2+}, Cu^{2+}, and Cr^{6+}, respectively, thereby indicating that EPSs play a major role in the removal of HIMs, corroborating the results of previous studies [7,10,71–73].

Figure 11a shows the FTIR spectra of the EPS filter cake before and after the adsorption of the various HIMs (i.e., EPS-Pb, EPS-Cu, and EPS-Cr). In addition, the spectra of the ENM and the EPS-contaminated ENM (EPS–ENM) are shown in Figure 11b. As indicated in Figure 11a, all the samples appeared to contain similar functional groups corresponding to polysaccharides, proteins, lipids, and nucleic acids, thereby indicating that the HIMs do not alter the molecular structure of EPSs upon their adsorption, and that the interaction mechanisms with the EPSs were similar for all three HIMs. As previously reported for similar experiments carried out using Pb^{2+} [10], the carboxylate residues of the EPSs also appeared to be bridged by the Cu^{2+} and Cr^{6+} ions.

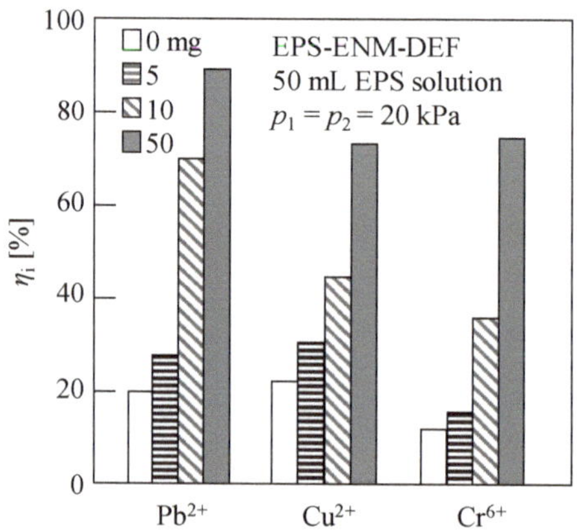

Figure 10. Removal efficiency (η_i) of the heavy-metal ions (HMIs) via dead-end filtration using the ENM-14 with an EPS filter cake on the surface (EPS–ENM–DEF). The evaluated data were obtained for the cumulative filtrate volume per unit membrane area, v = 9 cm. First stage: concentration and recovery filtration of 0, 0.1, 0.2, and 1.0 g/L EPS solutions (50 mL) at p_1 = 20 kPa to give EPS filter cakes with masses of 0, 5, 10, and 50 mg, respectively. Second stage: filtration of 10 μM solutions (180 mL) of the desired HIM (Pb^{2+}, Cu^{2+}, or Cr^{6+}, pH 6.2–6.7) at p_2 = 20 kPa.

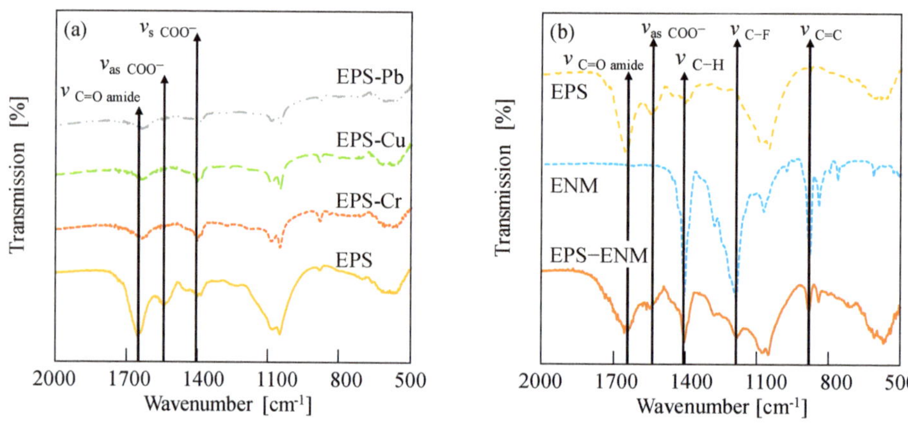

Figure 11. FTIR spectra for the various samples: (**a**) the EPS filter cake before and after the adsorption of HIMs to give EPS-Pb, EPS-Cu, and EPS-Cr; (**b**) the EPS, ENM, and EPS–ENM samples.

Figure 11b shows the FTIR spectra of the EPS, ENM, and EPS–ENM samples. More specifically, in the case of the EPS–ENM sample, three strong peaks were observed at 1402, 1167, and 882 cm^{-1}, which correspond to the C–H, C–F, and C–C moieties of PVDF ENM, respectively [69,74]. Meanwhile, two strong peaks were found at 1658 and 1539 cm^{-1}, which correspond to the C=O and COO^- groups of EPS, respectively. These results indicate that the EPS and ENM species are tightly bound to one another through electrostatic interactions attributed to the fluorine atoms of the PVDF ENM [75]. The electronegative nature of fluorine also accounts for the successful adsorption of small amounts of HIMs on the ENM specimen in the absence of the EPS filter cake. It is, therefore, expected that numerous mechanisms could be involved in the adsorption of HIMs on these EPS–ENMs, such as electrostatic attractions, complexation, ion exchange, and surface precipitation, thereby indicating the potential for preparing novel fibrous membranes with high HIM adsorption capacities in the future.

Furthermore, Figure 12 shows that the HIM removal efficiency varies during the EPS–ENM DEF process, wherein different results were obtained for the 10 and 50 mg EPS filter cakes. More specifically, as shown in Figure 12a, in the presence of a 50 mg EPS filter cake, the removal efficiencies of the three HIMs decreased with the filtration progress, wherein the greatest decline was observed for Cr^{6+}, as previously reported [10]. This can be accounted for by considering that, in the initial stage of EPS–ENM–DEF, the EPS filter cake contained numerous vacant adsorption sites; however, during the filtration process, these sites became occupied. In contrast, for the 10 mg EPS filter cake, after an initial decrease, the removal efficiencies of the three HIMs increased prior to becoming stable, as shown in Figure 12b. This was particularly apparent in the case of Cr^{6+}, wherein the initially low removal efficiency increased gradually but significantly, and this was attributed to the structural rearrangement of the EPS filter cake upon interaction with Cr^{6+} because of highly charged ions. Figure S5 also shows the HIM removal efficiencies during the EPS–ENM–DEF process on ENM-14 for the 5 and 0 mg EPS filter cakes. It should also be noted that the HIM removal efficiency is also likely influenced by a range of other factors, such as ion rejection by the cake or the ENM, ion adsorption by the ENM, and the filtration rate.

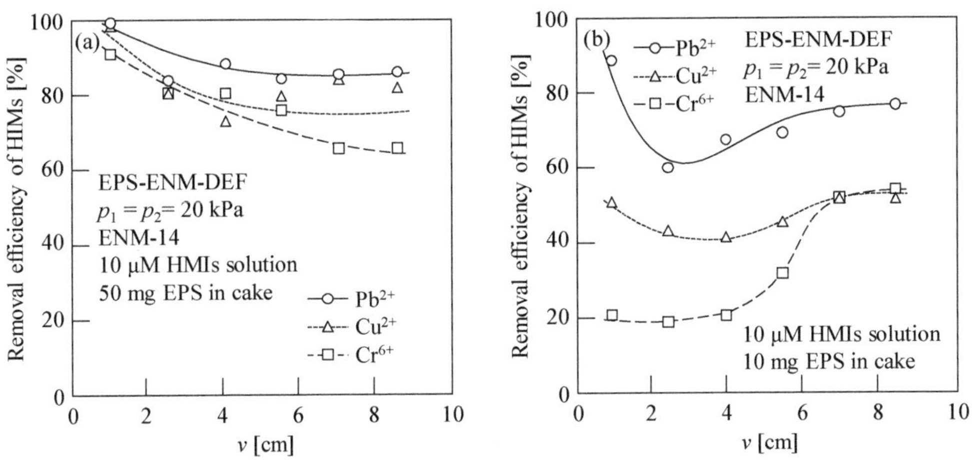

Figure 12. HIM removal efficiencies during the EPS–ENM–DEF process on ENM-14 for (**a**) 50 mg and (**b**) 10 mg EPS filter cakes. First stage: concentration and recovery filtration of 1.0 and 0.2 g/L EPS solutions (50 mL) at p_1 = 20 kPa to give EPS filter cakes with masses of 50 and 10 mg, respectively. Second stage: filtration of 10 μM solutions (180 mL) of the desired HIM (Pb^{2+}, Cu^{2+}, or Cr^{6+}, pH 6.2–6.7) at p_2 = 20 kPa. v is the cumulative filtrate volume per unit membrane area.

3.6. Advantages of Removing HIMs by EPS–ENM Filtration

As described above, the HMI pollution of natural water bodies and terrestrial ecosystems can be attributed to the persistence and bioaccumulation of such elements. Due to these factors, in addition to their toxic nature, HMIs post a great threat to vegetation, crops, and aquatic species [76]. The proposed two-stage EPS–ENM–DEF filtration process for HIM removal, therefore, appears to offer a number of advantages over traditional adsorption materials and processes. More specifically, the EPSs are derived from the excess sludge of wastewater treatment plants, and they have been proven to exhibit excellent adsorption properties [7,10,71–73], while also efficiently removing low concentrations of metal ions [77,78]; these factors greatly increase the value of such waste. Furthermore, in terms of complex pollutants, the adsorption capacities of EPSs toward certain multicomponent heavy-metal compositions have been found to be higher than those recorded for their individual single components [78,79], thereby confirming that EPSs appear suitable for dealing with complex water bodies.

Overall, our results indicate that these PVDF nanofiber membranes exhibit a high water flux that is retained during their repeated recycling, thereby permitting them to concentrate EPSs, with a recovery rate of up to 94.1% being achieved over five cycles. This was attributed to the high porosity and highly interconnected porous structure of the prepared ENMs. Table 1 shows a comparison of the typical parameters of various ENMs in the literature fabricated using PVDF, which have been reported over the past 5 years. These ENMs were optimized according to the types or proportions of solvents and modification materials (e.g., nanoparticles, carbon nanotubes, or graphene oxide) [27,31,49,50,52,80], thereby indicating that, in the future, it should be possible to develop superior ENMs that exhibit high water fluxes, strong mechanical properties, high reusability, and good antifouling properties through unique design and processing techniques. In particular, as indicated by the current study, ENMs exhibiting an electronegative surface can attract metal cations via electrostatic interactions, thereby improving their adsorption capabilities [81]. In our system, we found that both the EPS adsorbent and the HIMs were fixed on the ENM, which was easily separable from the clarified effluent, thereby significantly simplifying the post-treatment separation phase [56]. As a result, the application of ENMs to EPS recovery simplifies the practical application of EPSs to remove HIMs from wastewater systems. We, therefore, expect that EPS–ENMs will become an environmentally friendly class of materials for heavy-metal removal.

4. Conclusions

We herein reported the development of PVDF electrospun nanofiber membranes (ENMs) exhibiting a high water flux for the separation and concentration of extracellular polymeric substances (EPSs) extracted from excess sludge using the cation exchange resin (CER) method. The ENM prepared using a 14 wt.% loading of PVDF exhibited excellent properties, including a high water flux (376.8 $L/(m^2 \cdot h)$), which was 35.2 times greater than that achieved using a commercial 10 kDa membrane. In addition, an excellent EPS recovery rate of up to 94.1% was achieved using this ENM for the dead-end filtration (DEF) of a 1.0 g/L EPS solution at 20 kPa. Overall, the prepared ENMs displayed excellent mechanical strengths, antifouling properties, and high reusability, wherein their original performances were almost fully maintained after five recycles (with cleaning by ultrasonication between cycles). It was also found that the filtration pressure had little effect on EPS recovery rate and water flux. Subsequently, the removal of heavy-metal ions (HMIs) was achieved via a novel EPS–ENM–DEF process, wherein removal rates of 89.5%, 73.5%, and 74.6% were obtained for Pb^{2+}, Cu^{2+}, and Cr^{6+}, respectively. We, therefore, expect that, on the basis of our results and on the enhanced technology reported herein, further optimization will lead to the preparation of novel and excellent ENMs, e.g., different nanofiber materials and modified techniques, for the highly efficient concentration and recovery of EPSs extracted from excess sludge via CER. This system should also permit the combined removal of HIMs, which will be the focus of our future research in this area.

Supplementary Materials: The following supporting information can be downloaded at: https://www.mdpi.com/article/10.3390/membranes13010074/s1. Figure S1. Average EPS recovery rates obtained via the UV method ($\eta_{p\text{-}av}$) compared with those obtained via the weighing method ($\eta_{e\text{-}av}$); Figure S2. Typical images of the water contact angles of the ENMs; Figure S3. Tensile stress–strain curves of the fabricated ENMs; Figure S4. Typical size distributions of the colloids present in the EPS solutions; Figure S5. HIM removal efficiencies during the EPS–ENM–DEF process on ENM-14.

Author Contributions: Conceptualization, D.-Q.C.; methodology, D.-Q.C.; validation, D.-Q.C.; investigation, X.-D.L. and J.-L.H.; data curation, D.-Q.C.; writing—original draft preparation, D.-Q.C. and J.-L.H.; writing—review and editing, D.-Q.C.; supervision, D.-Q.C.; project administration, D.-Q.C., W.-Y.Z., X.-D.H., E.I. and N.K. All authors have read and agreed to the published version of the manuscript.

Funding: This research was partially funded by the Beijing Natural Science Foundation (8222043), the Beijing Nova Program (Z211100002121154), the Beijing High-level Innovation Team "Research of key technologies on traditional village protection and residential building function improvement", the BUCEA Postgraduate Innovation Project, and the Beijing Energy Conservation and Sustainable Urban and Rural Development Provincial and Ministry Co-Construction Collaboration Innovation Center.

Institutional Review Board Statement: Not applicable.

Data Availability Statement: Not applicable.

Conflicts of Interest: The authors declare no conflict of interest.

References

1. van Loosdrecht, M.C.M.; Brdjanovic, D. Anticipating the next century of wastewater treatment. *Science* **2014**, *344*, 1452–1453. [CrossRef]
2. Winkler, M.K.H.; van Loosdrecht, M.C.M. Intensifying existing urban wastewater. *Science* **2022**, *375*, 377–378. [CrossRef] [PubMed]
3. Kroiss, H. What is the potential for utilizing the resources in sludge? *Water Sci. Technol.* **2004**, *49*, 1–10. [CrossRef] [PubMed]
4. Gopinath, A.; Divyapriya, G.; Srivastava, V.; Laiju, A.R.; Nidheesh, P.V.; Suresh Kumar, M. Conversion of sewage sludge into biochar: A potential resource in water and wastewater treatment. *Environ. Res.* **2021**, *194*, 110656. [CrossRef] [PubMed]
5. Sarpong, G.; Gude, V.G. Codigestion and combined heat and power systems energize wastewater treatment plants—Analysis and case studies. *Renew. Sust. Energy. Rev.* **2021**, *144*, 110937. [CrossRef]
6. Cao, D.-Q.; Hao, X.-D.; Wang, Z.; Song, X.; Iritani, E.; Katagiri, N. Membrane recovery of alginate in an aqueous solution by the addition of calcium ions: Analyses of resistance reduction and fouling mechanism. *J. Membr. Sci.* **2017**, *535*, 312–321. [CrossRef]
7. Cao, D.-Q.; Song, X.; Fang, X.-M.; Yang, W.-Y.; Hao, X.-D.; Iritani, E.; Katagiri, N. Membrane filtration-based recovery of extracellular polymer substances from excess sludge and analysis of their heavy metal ion adsorption properties. *Chem. Eng. J.* **2018**, *354*, 866–874. [CrossRef]
8. Cao, D.-Q.; Song, X.; Hao, X.-D.; Yang, W.-Y.; Iritani, E.; Katagiri, N. Ca^{2+}-aided separation of polysaccharides and proteins by microfiltration: Implications for sludge processing. *Sep. Purif. Technol.* **2018**, *202*, 318–325. [CrossRef]
9. Cao, D.-Q.; Jin, J.-Y.; Wang, Q.-H.; Song, X.; Hao, X.-D.; Iritani, E.; Katagiri, N. Ultrafiltration recovery of alginate: Membrane fouling mitigation by multivalent metal ions and properties of recycled materials. *Chin. J. Chem. Eng.* **2020**, *28*, 2881–2889. [CrossRef]
10. Cao, D.-Q.; Wang, X.; Wang, Q.-H.; Fang, X.-M.; Jin, J.-Y.; Hao, X.-D.; Iritani, E.; Katagiri, N. Removal of heavy metal ions by ultrafiltration with recovery of extracellular polymer substances from excess sludge. *J. Membr. Sci.* **2020**, *606*, 118103. [CrossRef]
11. Cao, D.-Q.; Sun, X.-Z.; Yang, X.-X.; Hao, X.-D. News on alginate recovery by forward osmosis: Reverse solute diffusion is useful. *Chemosphere.* **2021**, *285*, 131483. [CrossRef]
12. Glińska, K.; Stüber, F.; Fabregat, A.; Giralt, J.; Font, J.; Bengoa, C. Moving municipal WWTP towards circular economy: Cellulose recovery from primary sludge with ionic liquid. *Resour. Conserv. Recycl.* **2020**, *154*, 104626. [CrossRef]
13. Felz, S.; Vermeulen, P.; van Loosdrecht, M.C.M.; Lin, Y.-M. Chemical characterization methods for the analysis of structural extracellular polymeric substances (EPS). *Water. Res.* **2019**, *157*, 201–208. [CrossRef]
14. Wei, L.; Li, J.; Xue, M.; Wang, S.; Li, Q.; Qin, K.; Jiang, J.; Ding, J.; Zhao, Q. Adsorption behaviors of Cu^{2+}, Zn^{2+} and Cd^{2+} onto proteins, humic acid, and polysaccharides extracted from sludge EPS: Sorption properties and mechanisms. *Bioresour. Technol.* **2019**, *291*, 121868. [CrossRef]
15. Kim, N.K.; Mao, N.; Lin, R.; Bhattacharyya, D.; van Loosdrecht, M.C.M.; Lin, Y. Flame retardant property of flax fabrics coated by extracellular polymeric substances recovered from both activated sludge and aerobic granular sludge. *Water Res.* **2020**, *170*, 115344. [CrossRef]
16. Kim, N.K.; Lin, R.; Bhattacharyya, D.; van Loosdrecht, M.C.M.; Lin, Y. Insight on how biopolymers recovered from aerobic granular wastewater sludge can reduce the flammability of synthetic polymers. *Sci. Total. Environ.* **2022**, *805*, 150434. [CrossRef]
17. More, T.T.; Yadav, J.S.S.; Yan, S.; Tyagi, R.D.; Surampalli, R.Y. Extracellular polymeric substances of bacteria and their potential environmental applications. *J. Environ. Manag.* **2014**, *144*, 1–25. [CrossRef]
18. Li, W.; Liu, M.; Siddique, M.S.; Graham, N.; Yu, W. Contribution of bacterial extracellular polymeric substances (EPS) in surface water purification. *Environ. Pollut.* **2021**, *280*, 116998. [CrossRef]
19. Siddharth, T.; Sridhar, P.; Vinila, V.; Tyagi, R.D. Environmental applications of microbial extracellular polymeric substance (EPS): A review. *J. Environ. Manag.* **2021**, *287*, 112307. [CrossRef]
20. Saxena, A.; Tripathi, B.P.; Kumar, M.; Shahi, V.K. Membrane-based techniques for the separation and purification of proteins: An overview. *Adv. Colloid Interface.* **2009**, *145*, 1–22. [CrossRef]
21. Li, Z.; Wang, K.; Zhang, Y. Eco-friendly separation and purification of soybean oligosaccharides via nanofiltration technology. *Sep. Purif. Technol.* **2018**, *53*, 777–785. [CrossRef]
22. Wu, D.; Feng, Q.; Li, M.; Wei, A.; Li, J.; Liu, C.; Xu, H.; Cheng, W. Preparation and protein separation properties of the porous polystyrene/ethylene–vinyl acetate copolymer blend nanofibers membranes. *ACS Omega* **2019**, *4*, 20152–20158. [CrossRef] [PubMed]

23. Teng, J.; Wu, M.; Chen, J.; Lin, H.; He, Y. Different fouling propensities of loosely and tightly bound extracellular polymeric substances (EPSs) and the related fouling mechanisms in a membrane bioreactor. *Chemosphere* **2020**, *255*, 126953. [CrossRef] [PubMed]
24. Homaeigohar, S.S.; Buhr, K.; Ebert, K. Polyethersulfone electrospun nanofibrous composite membrane for liquid filtration. *J. Membr. Sci.* **2010**, *365*, 68–77. [CrossRef]
25. Bassyouni, M.; Abdel-Aziz, M.H.; Zoromba, M.S.; Abdel-Hamid, S.M.S.; Drioli, E. A review of polymeric nanocomposite membranes for water purification. *J. Ind. Eng. Chem.* **2019**, *73*, 19–46. [CrossRef]
26. Ahmed, F.E.; Lalia, B.S.; Hashaikeh, R. A review on electrospinning for membrane fabrication: Challenges and applications. *Desalination* **2015**, *356*, 15–30. [CrossRef]
27. Gee, S.; Johnson, B.; Smith, A.L. Optimizing electrospinning parameters for piezoelectric PVDF nanofiber membranes. *J. Membr. Sci.* **2018**, *563*, 804–812. [CrossRef]
28. Abu-Obaid, S.; Aktij, S.A.; Tabe, S.; Sadrzadeh, M.; Farnood, R.R. Surfactant-modified adsorptive electrospun nanofiber membrane impregnated with akageneite for phosphorus recovery from wastewater. *J. Environ. Chem. Eng.* **2022**, *10*, 108786. [CrossRef]
29. Hu, G.; Zhang, X.; Liu, X.; Yu, J.; Ding, B. Electrospun nanofibers withstandable to high-temperature reactions: Synergistic effect of polymer relaxation and solvent removal. *Adv. Fiber Mater.* **2021**, *3*, 14–25. [CrossRef]
30. Meng, L.; Lv, Y.; Deng, P.; Li, N.; Huang, M.; Mansouri, J.; Chen, V. Novel PVDF membrane with sandwich structure for enhanced membrane distillation. *Chem. Eng. J.* **2021**, *415*, 128960. [CrossRef]
31. Venkatesh, K.; Arthanareeswaran, G.; Bose, A.C.; Kumarc, P.S.; Kweon, J. Diethylenetriaminepentaacetic acid-functionalized multi-walled carbon nanotubes/titanium oxide-PVDF nanofiber membrane for effective separation of oil/water emulsion. *Sep. Purif. Technol.* **2021**, *257*, 117926. [CrossRef]
32. Liang, Y.; Zhao, J.; Huang, Q.; Hu, P.; Xiao, C. PVDF fiber membrane with ordered porous structure via 3D printing near field electrospinning. *J. Membr. Sci.* **2021**, *618*, 118709. [CrossRef]
33. Lu, Z.; Zhang, B.; Gong, H.; Li, J. Fabrication of hierarchical porous poly (l-lactide) (PLLA) fibrous membrane by electrospinning. *Polymer* **2021**, *226*, 123797. [CrossRef]
34. Liao, Y.; Loh, C.H.; Tian, M.; Wang, R.; Fane, A.G. Progress in electrospun polymeric nanofibrous membranes for water treatment: Fabrication, modification and applications. *Prog. Polym. Sci.* **2018**, *77*, 69–94. [CrossRef]
35. Liu, F.; Wang, L.; Li, D.; Liu, Q.; Deng, B. Preparation and characterization of novel thin film composite nanofiltration membrane with PVDF tree-like nanofiber membrane as composite scaffold. *Mater. Design* **2020**, *196*, 109101. [CrossRef]
36. Zong, H.; Xia, X.; Liang, Y.; Dai, S.; Alsaedi, A.; Hayat, T.; Kong, F.; Pan, J.H. Designing function-oriented artificial nanomaterials and membranes via electrospinning and electrospraying techniques. *Mat. Sci. Eng. C* **2018**, *92*, 1075–1091. [CrossRef]
37. Li, W.; Chao, S.; Li, Y.; Bai, F.; Teng, Y.; Li, X.; Li, L.; Wang, C. Dual-layered composite nanofiber membrane with Cu-BTC-modified electrospun nanofibers and biopolymeric nanofibers for the removal of uremic toxins and its application in hemodialysis. *J. Membr. Sci.* **2022**, *642*, 119964. [CrossRef]
38. Liu, G.; Tsen, W.-C.; Jang, S.-C.; Hu, F.; Zhong, F.; Zhang, B.; Wang, J.; Liu, H.; Wang, G.; Wen, S.; et al. Composite membranes from quaternized chitosan reinforced with surface-functionalized PVDF electrospun nanofibers for alkaline direct methanol fuel cells. *J. Membr. Sci.* **2020**, *611*, 118242. [CrossRef]
39. Zhou, Z.; Lin, W.; Wu, X.-F. Electrospinning ultrathin continuous cellulose acetate fibers for high-flux water filtration. *Colloid Surface A Physicochem. Eng. Asp.* **2016**, *494*, 21–29. [CrossRef]
40. Yin, X.; Zhang, Z.; Ma, H.; Venkateswaran, S.; Hsiao, B.S. Ultra-fine electrospun nanofibrous membranes for multicomponent wastewater treatment: Filtration and adsorption. *Sep. Purif. Technol.* **2020**, *242*, 116794. [CrossRef]
41. Yadav, P.; Farnood, R.; Kumar, V. Superhydrophobic modification of electrospun nanofibrous Si@PVDF membranes for desalination application in vacuum membrane distillation. *Chemosphere* **2022**, *287*, 132092. [CrossRef] [PubMed]
42. Fu, Q.; Duan, C.; Yan, Z.; Si, Y.; Liu, L.; Yu, J.; Ding, B. Electrospun nanofibrous composite materials: A versatile platform for high efficiency protein adsorption and separation. *Compos. Commun.* **2018**, *8*, 92–100. [CrossRef]
43. Wang, Z.; Sahadevan, R.; Crandall, C.; Menkhaus, T.J.; Fong, H. Hot-pressed PAN/PVDF hybrid electrospun nanofiber membranes for ultrafiltration. *J. Membr. Sci.* **2020**, *611*, 118327. [CrossRef]
44. Han, J.-L.; Cao, D.-Q.; Hao, X.-D. Preparation of electrospun nanofibrous membrane for recovery and concentration of extracellular polymeric substances from excess sludge. In Proceedings of the Preprint of 52th Autumn Meeting, Online, 22–24 September 2021; The Society of Chemical Engineers: Okayama, Japan, 2021; p. PA354.
45. Yu, X.; Zhu, Y.; Cheng, C.; Zhang, T.; Wang, X.; Hsiao, B.S. Novel thin-film nanofibrous composite membranes containing directional toxin transport nanochannels for efficient and safe hemodialysis application. *J. Membr. Sci.* **2019**, *582*, 151–163. [CrossRef]
46. Ameduri, B. From vinylidene fluoride (VDF) to the applications of VDF-containing polymers and copolymers: Recent developments and future trends. *Chem. Rev.* **2009**, *109*, 6632–6686. [CrossRef] [PubMed]
47. Li, K.; Hou, D.; Fu, C.; Wang, K.; Wang, J. Fabrication of PVDF nanofibrous hydrophobic composite membranes reinforced with fabric substrates via electrospinning for membrane distillation desalination. *J. Environ. Sci.* **2019**, *75*, 277–288. [CrossRef]
48. Ren, J.; Woo, Y.C.; Yao, M.; Lim, S.; Tijing, L.D.; Shona, H.K. Nanoscale zero-valent iron (nZVI) immobilization onto graphene oxide (GO)-incorporated electrospun polyvinylidene fluoride (PVDF) nanofiber membrane for groundwater remediation via gravity-driven membrane filtration. *Sci. Total Environ.* **2019**, *688*, 787–796. [CrossRef]

49. Yang, Y.; Li, Y.; Cao, L.; Wang, Y.; Li, L.; Li, W. Electrospun PVDF-SiO_2 nanofibrous membranes with enhanced surface roughness for oil-water coalescence separation. *Sep. Purif. Technol.* **2021**, *269*, 118726. [CrossRef]
50. Nthunya, L.N.; Gutierrez, L.; Khumalo, N.; Derese, S.; Mamba, B.B.; Verliefde, A.R.; Mhlanga, S.D. Superhydrophobic PVDF nanofibre membranes coated with an organic fouling resistant hydrophilic active layer for direct-contact membrane distillation. *Colloid Surface A Physicochem. Eng. Asp.* **2019**, *575*, 363–372. [CrossRef]
51. Ke, G.; Jin, X.; Hu, H. Electrospun polyvinylidene fluoride/polyacrylonitrile composite fibers: Fabrication and characterization. *Iran. Polym. J.* **2020**, *29*, 37–46. [CrossRef]
52. Ahmadi, A.; Qanati, O.; Seyed Dorraji, M.S.; Rasoulifard, M.H.; Vatanpour, V. Investigation of antifouling performance a novel nanofibrous S-PVDF/PVDF and S-PVDF/PVDF/GO membranes against negatively charged oily foulants. *J. Membr. Sci.* **2017**, *536*, 86–97. [CrossRef]
53. Upadhyay, U.; Sreedhar, I.; Singh, S.A.; Patel, C.M.; Anitha, K.L. Recent advances in heavy metal removal by chitosan based adsorbents. *Carbohyd. Polym.* **2021**, *251*, 117000. [CrossRef]
54. Yang, Y.; Zhang, Y.; Zheng, H.; Zhang, B.; Zuo, Q.; Fan, K. Functionalized dual modification of covalent organic framework for efficient and rapid trace heavy metals removal from drinking water. *Chemosphere* **2022**, *290*, 133215. [CrossRef]
55. Phan, D.-N.; Khan, M.Q.; Nguyen, N.-T.; Phan, T.-T.; Ullah, A.; Khatri, M.; Kien, N.N.; Kim, I.-S. A review on the fabrication of several carbohydrate polymers into nanofibrous structures using electrospinning for removal of metal ions and dyes. *Carbohyd. Polym.* **2021**, *252*, 117175. [CrossRef]
56. Zhu, F.; Zheng, Y.-M.; Zhang, B.-G.; Dai, Y.-R. A critical review on the electrospun anofibrous membranes for the adsorption of heavy metals in water treatment. *J. Hazard. Mater.* **2021**, *401*, 123608. [CrossRef]
57. Efome, J.E.; Rana, D.; Matsuura, T.; Lan, C.Q. Metal-organic frameworks supported on nanofibers to remove heavy metals. *J. Mater. Chem. A* **2018**, *6*, 4550–4555. [CrossRef]
58. Pi, H.; Wang, R.; Ren, B.; Zhang, X.; Wu, J. Facile fabrication of multi-structured SiO_2@PVDF-HFP nanofibrous membranes for enhanced copper ions adsorption. *Polymers* **2018**, *10*, 1385. [CrossRef]
59. Ahmad, M.; Manzoor, K.; Ikram, S. Versatile nature of hetero-chitosan based derivatives as biodegradable adsorbent for heavy metal ions; a review. *Int. J. Biol. Macromol.* **2017**, *105*, 190–203. [CrossRef]
60. Wang, Y.; Ma, F.; Zhang, N.; Wei, X.; Yang, J.; Zhou, Z. Blend-electrospun poly (vinylidene fluoride)/polydopamine membranes: Self-polymerization of dopamine and the excellent adsorption/separation abilities. *J. Mater. Chem. A* **2017**, *5*, 14430–14443.
61. Ma, F.-F.; Zhang, D.; Zhang, N.; Huang, T.; Wang, Y. Polydopamine-assisted deposition of polypyrrole on electrospun poly (vinylidene fluoride) nanofibers for bidirectional removal of cation and anion dyes. *Chem. Eng. J.* **2018**, *354*, 432–444. [CrossRef]
62. Managheb, M.; Zarghami, S.; Mohammadi, T.; Asadi, A.A.; Sahebi, S. Enhanced dynamic Cu(II) ion removal using hot-pressed chitosan /poly (vinyl alcohol) electrospun nanofibrous affinity membrane(ENAM). *Process Saf. Environ. Prot.* **2021**, *146*, 329–337. [CrossRef]
63. Chen, H.; Huang, M.; Liu, Y.; Meng, L.; Ma, M. Functionalized electrospun nanofiber membranes for water treatment: A review. *Sci. Total Environ.* **2020**, *739*, 139944. [CrossRef] [PubMed]
64. Cao, D.-Q.; Iritani, E.; Katagiri, N. Properties of filter cake formed during dead-end microfiltration of O/W emulsion. *J. Chem. Eng. Jpn.* **2013**, *46*, 593–600. [CrossRef]
65. Iritani, E.; Katagiri, N.; Takenaka, T.; Yamashita, Y. Membrane pore blocking during cake formation in constant pressure and constant flux dead-end microfiltration of very dilute colloids. *Chem. Eng. Sci.* **2015**, *122*, 465–473. [CrossRef]
66. Iritani, E.; Katagiri, N. Developments of blocking filtration model in membrane filtration. *KONA Powder Part. J.* **2016**, *33*, 179–202. [CrossRef]
67. Ferraris, S.; Cazzola, M.; Peretti, V.; Stella, B.; Spriano, S. Zeta potential measurements on solid surfaces for in vitro biomaterials testing: Surface charge, reactivity upon contact with fluids and protein absorption. *Front. Bioeng. Biotechnol.* **2018**, *6*, 60. [CrossRef]
68. Liu, X.; Ma, H.; Hsiao, B.S. Interpenetrating nanofibrous composite membranes for water purification. *ACS Appl. Nano Mater.* **2019**, *2*, 3606–3614. [CrossRef]
69. Park, M.J.; Gonzales, R.R.; Abdel-Wahab, A.; Phuntsho, S.; Shon, H.K. Hydrophilic polyvinyl alcohol coating on hydrophobic electrospun nanofiber membrane for high performance thin film composite forward osmosis membrane. *Desalination* **2018**, *426*, 50–59. [CrossRef]
70. Khayet, M.; García-Payo, C.; Matsuura, T. Superhydrophobic nanofibers electrospun by surface segregating fluorinated amphiphilic additive for membrane distillation. *J. Membr. Sci.* **2019**, *588*, 117215. [CrossRef]
71. Sheng, G.-P.; Xu, J.; Luo, H.-W.; Li, W.-W.; Li, W.-H.; Yu, H.-Q.; Xie, Z.; Wei, S.-Q.; Hu, F.-C. Thermodynamic analysis on the binding of heavy metals onto extracellular polymeric substances (EPS) of activated sludge. *Water Res.* **2013**, *47*, 607–614. [CrossRef]
72. Wang, Y.; Qin, J.; Zhou, S.; Lin, X.; Ye, L.; Song, C.; Yan, Y. Identification of the function of extracellular polymeric substances (EPS) in denitrifying phosphorus removal sludge in the presence of copper ion. *Water Res.* **2015**, *73*, 252–264. [CrossRef]
73. Nkoh, J.N.; Lu, H.-L.; Pan, X.-Y.; Dong, G.; Kamran, M.A.; Xu, R.-K. Effects of extracellular polymeric substances of Pseudomonas fluorescens, citrate, and oxalate on Pb sorption by an acidic Ultisol. *Ecotoxicol. Environ. Saf.* **2019**, *171*, 790–797. [CrossRef]
74. Zhang, C.; Mao, H.; Cui, R.; Zhang, X.; Yang, J.; Ji, J.; Zhou, X. Electrospinning preparation, energetic characteristics and reaction mechanism of corrosion-resistant Si@PVDF nanostructured energetic films. *Combust. Flame* **2022**, *237*, 111887. [CrossRef]

75. Valluri, S.K.; Schoenitz, M.; Dreizin, E. Fluorine-containing oxidizers for metal fuels in energetic formulations. *Def. Technol.* **2019**, *15*, 1–22. [CrossRef]
76. Liu, R.; Jiang, W.; Li, F.; Pan, Y.; Wang, C.; Tian, H. Occurrence, partition, and risk of seven heavy metals in sediments, seawater, and organisms from the eastern sea area of Shandong Peninsula, Yellow Sea, China. *J. Environ. Manag.* **2021**, *279*, 111771. [CrossRef]
77. Zhang, Z.-Z.; Deng, R.; Cheng, Y.-F.; Zhou, Y.-H.; Buayi, X.; Zhang, X.; Wang, H.-Z.; Jin, R.-C. Behavior and fate of copper ions in an anammox granular sludge reactor and strategies for remediation. *J. Hazard. Mater.* **2015**, *300*, 838–846. [CrossRef]
78. Pagliaccia, B.; Carretti, E.; Severi, M.; Berti, D.; Lubello, C.; Lotti, T. Heavy metal biosorption by Extracellular Polymeric Substances (EPS) recovered from anammox granular sludge. *J. Hazard. Mater.* **2022**, *424*, 126661. [CrossRef]
79. Sun, X.-F.; Wang, S.G.; Zhang, X.-M.; Paul Chen, J.; Li, X.-M.; Gao, B.-Y.; Ma, Y. Spectroscopic study of Zn^{2+} and Co^{2+} binding to extracellular polymeric substances (EPS) from aerobic granules. *J. Colloid Interface Sci.* **2009**, *335*, 11–17. [CrossRef]
80. Ma, W.; Jiang, Z.; Lu, T.; Xiong, R.; Huang, C. Lightweight, elastic and superhydrophobic multifunctional nanofibrous aerogel for self-cleaning, oil/water separation and pressure sensing. *Chem. Eng. J.* **2022**, *430*, 132989. [CrossRef]
81. Wang, B.-B.; Liu, X.-T.; Chen, J.-M.; Peng, D.-C.; He, F. Composition and functional group characterization of extracellular polymeric substances (EPS) in activated sludge: The impacts of polymerization degree of proteinaceous substrates. *Water Res.* **2018**, *129*, 133–142. [CrossRef]

membranes

Article

Calcium Alginate Production through Forward Osmosis with Reverse Solute Diffusion and Mechanism Analysis

Da-Qi Cao [1,*], Kai Tang [1], Wen-Yu Zhang [2], Cheng Chang [3], Jia-Lin Han [1], Feng Tian [1] and Xiao-Di Hao [1]

1 Sino-Dutch R&D Centre for Future Wastewater Treatment Technologies/Key Laboratory of Urban Stormwater System and Water Environment, Beijing University of Civil Engineering and Architecture, Beijing 100044, China
2 Institute of Soil Environment and Pollution Remediation, Beijing Municipal Research Institute of Environmental Protection, Beijing 100037, China
3 Institute of Chemical Engineering, Chemical and Process Engineering, School of Engineering and Physical Sciences, Heriot-Watt University, Edinburgh EH14 4AS, UK
* Correspondence: caodaqi18@163.com or caodaqi@bucea.edu.cn; Tel.: +86-10-6832-2123

Abstract: Calcium alginate (Ca-Alg) is a novel target product for recovering alginate from aerobic granular sludge. A novel Ca-Alg production method was proposed herein where Ca-Alg was formed in a sodium alginate (SA) feed solution (FS) and concentrated via forward osmosis (FO) with Ca^{2+} reverse osmosis using a draw solution of $CaCl_2$. An abnormal reverse solute diffusion was observed, with the average reverse solute flux (RSF) decreasing with increasing $CaCl_2$ concentrations, while the average RSF increased with increasing alginate concentrations. The RSF of Ca^{2+} in FS decreased continuously as the FO progressed, using 1.0 g/L SA as the FS, while it increased initially and later decreased using 2.0 and 3.0 g/L SA as the FS. These results were attributed to the Ca-Alg recovery production (CARP) formed on the FO membrane surface on the feed side, and the percentage of Ca^{2+} in CARP to total Ca^{2+} reverse osmosis reached 36.28%. Scanning electron microscopy and energy dispersive spectroscopy also verified CARP existence and its Ca^{2+} content. The thin film composite FO membrane with a supporting polysulfone electrospinning nanofiber membrane layer showed high water flux and RSF of Ca^{2+}, which was proposed as a novel FO membrane for Ca-Alg production via the FO process with Ca^{2+} reverse diffusion. Four mechanisms including molecular sieve role, electrification of colloids, osmotic pressure of ions in CARP, and FO membrane structure were proposed to control the Ca-Alg production. Thus, the results provide further insights into Ca-Alg production via FO along with Ca^{2+} reverse osmosis.

Keywords: forward osmosis; calcium alginate; recovery; reverse solute diffusion; thin film composite forward osmosis membrane; electrospinning nanofiber membrane

Citation: Cao, D.-Q.; Tang, K.; Zhang, W.-Y.; Chang, C.; Han, J.-L.; Tian, F.; Hao, X.-D. Calcium Alginate Production through Forward Osmosis with Reverse Solute Diffusion and Mechanism Analysis. *Membranes* **2023**, *13*, 207. https://doi.org/10.3390/membranes13020207

Academic Editor: Sangho Lee

Received: 7 December 2022
Revised: 26 January 2023
Accepted: 6 February 2023
Published: 8 February 2023

1. Introduction

Alginate is a highly valued polysaccharide and is mainly obtained from seaweed; however, large amounts of fresh water and energy are consumed in the production process [1–3]. Recently, alginate was found to account for 15–25% of the total sludge dry weight in aerobic granular sludge (AGS) [2,4]. Therefore, applications of specific granulation of AGS [5,6] were proposed to achieve the targeted alginate production instead of conventional industrial methods that extracted alginate from seaweed [7–10], thereby contributing to a circular economy and society.

Sodium alginate (SA) is the most commercially produced form of alginate [1,3]. Recently, calcium-alginate (Ca-Alg) that can be prepared from SA by metathesis with appropriate calcium salts has been considered as a novel recyclable alginate material because of its high water-absorbing properties, biocompatibility, and its 3D structure that mimics the extracellular matrix [7,11,12]. These Ca-Alg characteristics promote healing of chronic wounds, thus enabling the widespread use of Ca-Alg in wound dressings [13]. Ca-Alg beads, well-known for their special adsorption ability, are used in producing dyes [14], fungicides like difenoconazole, and insecticides such as nitenpyram [15]. As a biodegradable biomaterial or composite adsorbent, Ca-Alg demonstrates excellent removal/recovery performance for toxic metals (Cu^{2+} and Ni^{2+} ions) [16], valuable trace metals, rare earth ions (e.g., Li, Sr, and La) [17], U (VI) ions [18], and ammonia nitrogen and phosphate [19]. Using Ca-Alg in AGS as a coagulant can retain more biomass and improve wastewater treatment efficiency [20]. Additionally, Ca-Alg enables a good biological microecological environment that can encapsulate microscale zero-valent iron [21], diatom biosilica, and Bacillus subtilis [22] for sewage treatment. Ca-Alg hydrogels can also be used to prepare composite filtration membranes with nanofiber support layers that can be used for molecule/ion separation owing to their good anti-pollution performance [23,24]. Furthermore, Ca-Alg capsules can be used as building materials that can gradually release healing agents without being ruptured under cyclic traffic loading, and are an economically viable, environmentally friendly, and promising self-healing mix material for asphalt pavement [25–27].

The alginate solution extracted from AGS is close to 100% water content [3,7–10,28], which necessitates increased energy consumption when direct drying methods such as spray- and freeze-drying are used. Membrane separation is an effective method for alginate concentration without adding any chemical reagent, thus avoiding secondary pollution [3,10]. During membrane separation methods, forward osmosis (FO), a low-carbon and pollution-free method that does not require external power, has been used recently for determining alginate concentrations [8–10]. However, some amount of draw solute from the draw solution (DS) side can diffuse reversely into the feed solution (FS) side because both the active and support layers have a two-way transport ability. The reverse solute diffusion (RSD) reduces the osmotic driving pressure, contaminates the FS, and increases the consumption cost of the draw solute [10,29]. Therefore, RSD in the DS is a recurring problem in FO technology that requires urgent attention [30,31]. Our previous study reported the beneficial effects of RSD in the FO concentration of alginate, where RSD was not deemed as a drawback in the DS [9,10]. However, this phenomenon requires further investigation. Ca-Alg, which is a recycled material, as mentioned previously, can be formed on the FO membrane on the feed side through the reverse infiltration of Ca^{2+} in the DS.

Other studies have also reported the benefits of RSD. For example, Mg^{2+} reverse osmosis was used to form high-purity struvite on the membrane surface, achieving >50% water recovery, >93% ammonium nitrogen removal, and >99% phosphate recovery [32]. RSD can positively influence the system performance using $NaHCO_3$ solution as a DS (catholyte) in an osmotic microbial fuel cell through self-buffering with reverse-fluxed sodium bicarbonate [33]. Reverse osmosis using ammonia nitrogen could release extracellular polymers in the microalgae on the FS side during FO dewatering of algae using urine as the DS [34].

Furthermore, a novel FO membrane that can control the reverse solute flux (RSF) of Ca^{2+} and enable a high water flux must be fabricated. However, the conventional commercial FO membranes are manufactured to avoid a trade-off between membrane permeability and selectivity [35], in which the FO membranes are designed to minimize RSF. The nanofiber structure resulting from electrostatic spinning has many excellent characteristics, such as high porosity and permeability, large specific surface area per unit mass, and small inter-fiber pores [36]. A porous nanofiber support layer can reduce the inner concentration polarization and increase water flux by 7- to 8-fold. As such, its use has been proposed for thin film composite FO membranes [37–40]. Typically, electrostatic spinning materials such as polyethylene terephthalate (PET) [38], polyvinylidene fluoride (PVDF) [41], and polysulfone (PSf) [42] are used. Alternatively, the RSF of the draw solute and the water flux can be controlled by adjusting the preparation parameters of interfacial polymerization (IP) such as the ratio of m-phenylenediamine (MPD) and trimesoyl chloride (TMC) in the active polyamide layer [41].

This study aimed to establish a novel Ca-Alg production technique and analyze its mechanisms and influence factors on FO with RSD. The influence of the DS and FS concentration on the reverse solute flux (RSF) and water flux and the content of Ca^{2+} in recycled materials (Ca-Alg recovery production, CARP) formed on the FO membrane on the FS side were investigated using $CaCl_2$ as the draw solute and SA as the feed in FO. Scanning electron microscopy (SEM) and energy dispersion spectroscopy (EDS) were used to analyze the CARP. The thin film composite (TFC) FO membranes with electrospinning nanofibers as a porous support layer and IP polyamide as an active layer were fabricated, and their alginate recovery properties, such as specific RSF and water flux, were compared to those of commercial FO membranes. The basic characteristics of Ca-Alg concentration recovery in the FO process using Ca^{2+} reverse osmosis were obtained.

2. Materials and Methods

2.1. Materials

SA (molecular weight (MW) = 120–190 kDa) was obtained from Sigma-Aldrich (St. Louis, MO, USA). $CaCl_2$ (analytical grade) was purchased from the Chinese Medicine Group Chemical Reagent Co., Ltd., Beijing, China. HCl and HNO_3 were sourced from Beijing Chemical Reagent Research Institute Co., Ltd., Beijing, China. Polyvinylidene fluoride (PVDF, HSV900, MW $\approx$ 1000 kDa, Arkema Chemical Co., Ltd., Shanghai, China) and polysulfone (PSf) (MW $\approx$ 22,000, Merck Co., Ltd., Kenneworth, Rahway, NJ, USA) were used as electrospinning polymers, while N,N-Dimethylformamide (DMF, analytical reagent, 99.5%, Shanghai Yien Chemical Technology Co., Ltd., Shanghai, China) and 1-methyl-2-pyrrolidinone (NMP, 99.5%, Merck) were used as solvents during membrane substrate fabrication. Additionally, 1,3,5-benzenetricarbonyl trichloride (TMC, >98%, Macklin Biochemical Technology Co., Ltd., Shanghai, China), n-hexane (C_6H_6, >99%, Macklin Biochemical Technology Co., Ltd., Shanghai, China), and m-phenylenediamine (MPD, >99%, Merck Co., Ltd., Kenneworth, USA) were used to prepare the polyamide (PA) layer of membranes. Cellulose triacetate, with an embedded polyester screen support (CTA-ES) FO membrane (Hydration Technology Innovations Corp., Albany, USA), was used. Ultrapure water (resistivity $\geq$ 18.2 MΩ) was obtained by purifying tap water using an Arium Comfort II ultrapure water system for laboratory use (Sartorius Corp., Göttingen, Germany). Further, 0.22 μm polyethersulfone disposable filter membrane used in the analytical methods was purchased from Jinteng Experimental Equipment Co., Ltd., Tianjin, China.

2.2. Experimental Setup and Technique

The FO experimental setup consisted of the following elements: an FO filter unit (CF042A-FO, Sterlitech Corp., Kent, Auburn, WA, USA), with an FO membrane having an effective area of 42.0 cm^2, two variable speed gear pumps for recirculating the FS and DS (WT3000-1JA, Baoding Lange Pump Corp., Baoding, China), an electronic balance for measuring the increase in the DS weight (BSA3202S, Sartorius Corp., Gottingen, Germany), a magnetic stirrer (FA2004, Shanghai Shunyu Hengping Scientific Instrument Corp., Shanghai, China) to constantly stir the FS, a computer for data recording, and connecting rubber pipes [10,43].

SA was dissolved and stirred in ultrapure deionized (DI) water at 24 °C for 12 h to obtain the SA solution (1.0–3.0 g/L). $CaCl_2$ was added to ultrapure DI water and dissolved by stirring for 3 h. This solution was used as the corresponding DS of $CaCl_2$ (1.0–3.0 M). The FO experiment was conducted using 500 mL FS and 1500 mL DS, based on sweep model (flow rates, u = 2.5 cm/s along opposite directions on both sides of the membrane) [43] or dead-end model (u = 3.0 cm/s for the DS side and u = 0 cm/s for the FS side) [44]. The cell with a 5 cm height internal cavity was used on the FS side for the dead-end model. The forward osmosis time of 5 h was used for convenience at present and 10 mL of the FS solution was removed every 1 h and used as a sample. The FS volume was calculated according to the increasing DS mass. Notably, the osmotic pressure of the DS can be considered as a constant in the FO process; therefore, it was not tested.

2.3. Preparation of TFC FO Membranes

The electrospinning nanofiber membrane (ENM) substrate was fabricated using an electrospinning machine (ET2535, Yongkang Leye Co., Ltd., Beijing, China). First, electrospinning polymer made of PSf or PVDF (20 wt%) was dissolved in a DMF and NMP solution (volume ratio of 7:3) at a stirring speed of 300 rpm and temperature of 60 °C for 10 h in a magnetic stirring constant temperature water bath (Gaode instrument manufacturing Co., Ltd., Changzhou, China). The polymer solution was then transferred into a plastic syringe. To collect the solution on an aluminum-covered drum, we operated a syringe pump (G20, ID 0.60 mm) at a rate of 0.1 mm/min under high voltage (12.5 kV), placed 16 cm from the drum. The temperature and humidity inside the electrospinning chamber were set to 25 °C and 40%, respectively, in which the electrospinning nanofiber support layer was prepared. Afterwards, the support layer was soaked in an anhydrous ethanol solution for 2 h to improve its hydrophilicity and then stored in ultrapure water.

As pretreatment, the substrates were soaked in ethanol for 2 h prior to interfacial polymerization (IP) for fabrication of FO membranes. Then, a 3.4 wt% MPD and ultrapure water solution was poured onto the nanofiber substrate surface for 5 min, removing the excess solution through nitrogen blowing. Then, the MDP saturated substrate was submerged in a 0.075 wt% TMC solution for 2 min to allow the formation of a PA layer. Finally, the obtained membranes were cured in an oven at 45 °C for 10 min and then stored in ultrapure water for later use, denoted as PSf or PVDF ENM-TFC.

2.4. Analytical Methods

A sample solution of 4 mL was taken from the feed and mixed with 0.2 mL HCl solution (50%), and the resultant mixed solution was centrifuged at 4000 rpm for 20 min. Subsequently, the supernatant was collected and added to 0.2 mL HNO_3 solution (1%) to remove Ca^{2+}. Later, the mixed solution was filtered through a 0.22 μm pore diameter membrane and the Ca^{2+} concentration in the filtrate was used to determine the Ca^{2+} concentration in the sample solution. The metal ion concentrations in the aqueous solutions were measured using an inductively coupled plasma spectrometer (ICAP 7000 Series, Thermo Fisher Scientific Co., Ltd., Waltham, MA, USA). Ultrapure water was used as a blank sample with a metal ion concentration of 0 mM.

The CARP after the FO progressed was cut into a 1 cm × 1 cm small piece to observe its surface morphology. Some CARP samples were dried for 48 h by vacuum freeze-drying (FD-1A-50, Beijing Boyikang Experimental Instrument Corp., Beijing, China) and then freeze-dried in liquid nitrogen to prepare their cross-sections. All specimens were sputter-coated with gold for 10 min and then observed under a scanning electron microscope (SEM) (G300, ZEISS Corp., Oberkohen, Germany) operating at 200–7000 V under low vacuum conditions. Elemental analysis of the interface was performed by an energy-dispersive X-ray spectrometer (EDS, Oxford Xplore30) attached to the SEM. Five different points of each sample were observed and digital pictures were subsequently taken.

The pore sizes of the ENMs were determined through mercury porosity experiments (AutoPore Iv 9510, Micromeritics Instrument Co., Ltd., Shanghai, China) and the water contact angles of the ENMs were measured using a contact angle meter (JC2000D4M, Zhongchen Digital Technology Equipment Co., Ltd., Shanghai, China). Finally, the mechanical properties of each sample (shaped into strips of 2 cm × 1 cm) were tested using an electronic universal testing machine (non-metallic direction) (Inspekt table blue 5 kN, Hegewald & Peschke, Dresden, Germany) at a distance of 20 cm and a tensile rate of 2 mm/min.

2.5. *Evaluation of Forward Osmosis Performance*

Figure 1 shows the mass transfer diagram in Ca-Alg production via FO using reverse solute diffusion. Herein, the volume of CARP and the concentration change of Ca^{2+} in DS are negligible. The water flux (J_w) passing through the FO membrane per unit membrane area per unit time at time t was calculated by the numerical differentiation of the volume versus time data [10,43], and is expressed as follows:

$$J_w = \frac{dV_{DS}}{A_m dt} \quad (1)$$

where V_{DS} is the DS volume, A_m is the effective area of the FO membrane, and t is the FO time.

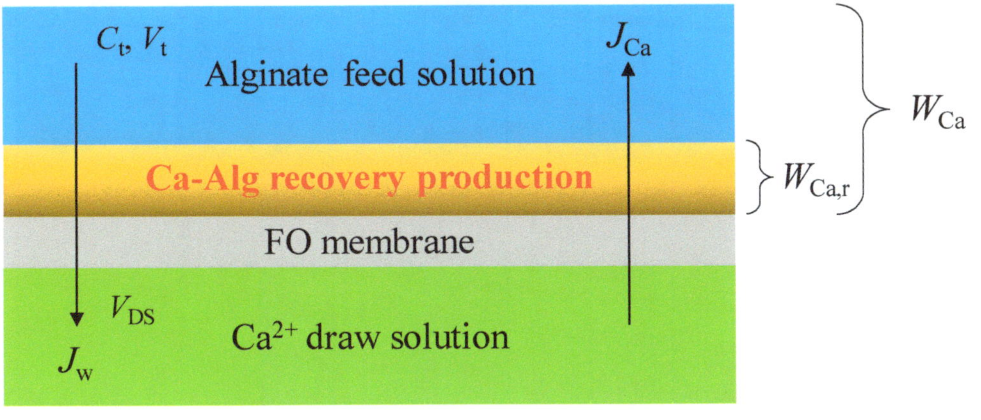

Figure 1. Mass transfer diagram in calcium alginate production via forward osmosis (FO) using reverse solute diffusion. Here, C_t is the Ca^{2+} concentration in the feed solution (FS) at time t; V_t and V_{DS} are the volumes of the FS and the draw solution (DS) at time t, respectively; J_w is the water flux passing through the FO membrane per unit membrane area per unit time at time t; J_{Ca} is the reverse solute flux (RSF) of Ca^{2+} without considering Ca^{2+} in the CARP passing through the FO membrane per unit membrane area per unit time at time t; W_{Ca} is the total content of reverse osmosis Ca^{2+} including that in the CARP formed on the FO membrane at time t; $W_{Ca,r}$ is the Ca^{2+} content in CARP formed on the FO membrane at time t.

The total average RSF of Ca^{2+} ($J_{Ca,av}$) in Figure 1, considering Ca^{2+} in the CARP passing through the FO membrane per unit membrane area from time $t = 0$ to t, is calculated as follows:

$$J_{Ca,\ av} = \frac{W_{Ca}}{A_m t} = \frac{W_{Ca,r} + C_t V_t}{A_m t} \quad (2)$$

where W_{Ca} is the total content of reverse osmosis Ca^{2+} including that in the CARP formed on the FO membrane at time t; $W_{Ca,r}$ is the Ca^{2+} content in CARP formed on the FO membrane at time t; and C_t and V_t are the concentration of Ca^{2+} in the FS without considering Ca^{2+} in the CARP and the FS volume at time t, respectively.

The RSF of Ca^{2+}, J_{Ca}, without considering that in the CARP passing through the FO membrane per unit membrane area per unit time at time t, is calculated as follows:

$$J_{Ca} = \frac{C_{t+t_0} V_{t+t_0} - C_t V_t}{A_m t_0} \quad (3)$$

where C_{t+t_0} is the concentration of Ca^{2+} in FS at time (t + t_0), V_{t+t_0} is the FS volume at time (t + t_0), and t_0 is the time interval between two consecutive samplings.

3. Results and Discussion

3.1. Reverse Solute Diffusion of Calcium Ion for Producing Ca-Alg in Forward Osmosis

The concentrated Ca-Alg was formed on the FO membrane on the feed side as a result of Ca^{2+} reverse osmosis in the DS when the $CaCl_2$ solution was used as the DS [10]. As this study proposed a novel Ca-Alg production method, in which Ca-Alg was formed on the FO membrane surface by the interaction between SA and reverse osmosis Ca^{2+} in FO process, the RSF is more helpful than the ratio of RSF to water flux. The Ca^{2+} RSF for various concentrations of FS and DS in the FO process was evaluated to comprehensively discuss the reverse solute diffusion. After 5 h of FO experimentation, the Ca^{2+} content in the FS containing Ca^{2+} in the CARP formed on the FO membrane was measured and the average RSF ($J_{Ca,av}$) was calculated using Equation (2). As shown in Figure 2a, an abnormal negative correlation of the average RSF was observed with the increasing $CaCl_2$ concentration; additionally, the average RSF decreased with the increase in the Ca^{2+} concentration when 1.0 g/L SA solution was used as the FS; in contrast, the average RSF increased with increasing $CaCl_2$ concentrations when ultrapure water was used as the FS. Furthermore, the average Ca^{2+} RSF when SA solution was used as the FS was lower than that when ultrapure water was used as the FS, because of the larger resistance of RSD caused by the CARP formed on the FO membrane surface.

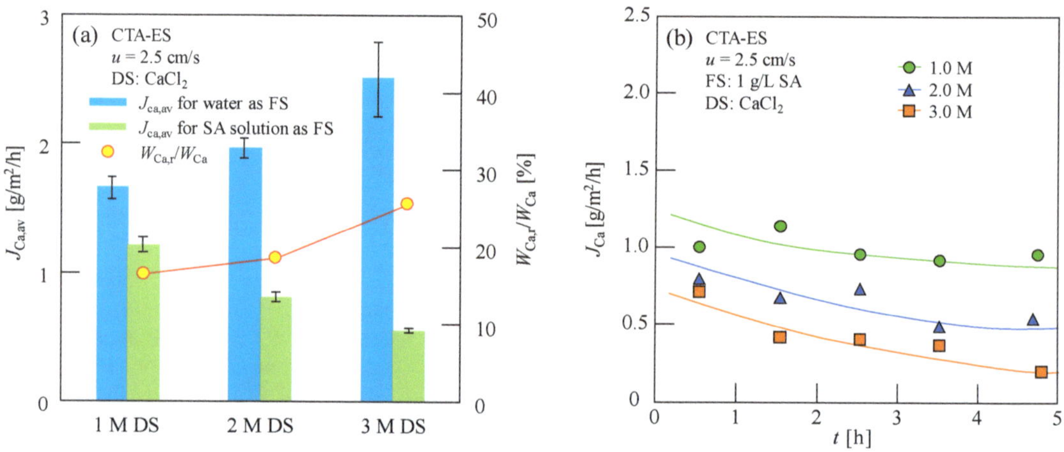

Figure 2. (**a**) Average reverse solute flux (RSF) of Ca^{2+} ($J_{Ca,av}$) for ultrapure water and sodium alginate (SA) solution as feed solution (FS) and percentage of Ca^{2+} content in Ca-Alg recovery production (CARP) formed on the forward osmosis (FO) membrane ($W_{Ca,r}$) to total Ca^{2+} content (W_{Ca}) in the FS containing Ca^{2+} in the CARP, $W_{Ca,r}/W_{Ca}$, for SA solution as the FS in FO with various $CaCl_2$ solutions as the draw solution (DS); (**b**) RSF of Ca^{2+} without considering Ca^{2+} in the CARP (J_{Ca}) with the progress of FO. Here, 1.0 g/L SA solution and 1–3 M $CaCl_2$ solution were used as the FS and DS, respectively; for the flow rates of 2.5 cm/s, u along opposite directions on both sides of the membrane was used; CTA-ES, cellulose triacetate with embedded polyester screen support.

Changes in the RSF during the FO process were investigated to further explore the abnormal results obtained for the negative correlation of the average RSF. The RSF of Ca^{2+} without Ca^{2+} in the CARP for 0–5 h as the FO progressed for 1.0 g/L SA solution, used as the FS with various DS concentrations, is shown in Figure 2b. Generally, the RSF value of Ca^{2+} remains constant during the FO process using ultrapure water as the FS; however, as shown in Figure 2b, the RSF value of Ca^{2+} decreased gradually in the FO process when the SA solution was used as the FS. It should be noted that the lines in Figure 2b simply indicate the variation trends of J_{Ca} with FO progress for a clear understanding, because of the lack of a specific functional relationship between J_{Ca} and t in this study. Furthermore, the initial RSF value of Ca^{2+} (obtained by extrapolation at $t = 0$), as shown in Figure 2b, also differed from the average RSF value of Ca^{2+} in the FO that used ultrapure water as the FS (Table 1), although no membrane fouling was observed at the beginning of the FO.

Table 1. Characteristic values in forward osmosis (FO) recovery of Ca-Alg at 5 h of FO. Herein, commercial CTA-ES FO membrane, 1–3 M $CaCl_2$ draw solution (DS), 0–3 g/L sodium alginate (SA) feed solution (FS), and 2.5 cm/s flow rates along opposite directions on both sides of the membrane were used.

Conc. of DS [M]	Conc. of SA [g/L]	W_{Ca} [a] [mg]	$W_{Ca,r}$ [b] [mg]	$W_{Ca,r}/W_{Ca}$ [c] [%]	W_{Ca}/W_{SA} [d] [%]	$J_{Ca,av}$ [e] [g/m²/h]	$J_{w,av}$ [f] [L/m²/h]
1	0	34.85	- [g]	-	-	1.66 ± 0.12	10.66 ± 0.28
	1	25.55	4.19	16.40	5.11	1.22 ± 0.14	6.63 ± 0.23
2	0	41.43	-	-		1.97 ± 0.11	15.52 ± 0.16
	1	16.99	3.13	18.42	3.40	0.81 ± 0.09	9.30 ± 0.09
	2	35.43	7.92	22.35	3.54	1.69 ± 0.07	7.45 ± 0.13
	3	44.30	16.07	36.28	2.95	2.11 ± 0.05	7.56 ± 0.15
3	0	52.71	-	-	-	2.51 ± 0.21	19.29 ± 0.32
	1	11.59	2.97	25.63	2.32	0.55 ± 0.02	9.58 ± 0.18

[a] Total content of reverse osmosis Ca^{2+} containing that in the Ca-Alg recovery production (CARP) formed on the FO membrane; [b] content of Ca^{2+} in the CARP; [c] percentage content of Ca^{2+} in the CARP to total osmosis content of Ca^{2+}; [d] content rate of total Ca^{2+} reverse osmosis to SA in the FS; [e] average flux of total Ca^{2+} reverse osmosis; [f] average water flux; [g] none.

The average Ca^{2+} content of reverse osmosis ($J_{Ca,av}$) after 5 h of FO, with 1.0–3.0 g/L SA solution as the FS and 2.0 M $CaCl_2$ solution as the DS, is shown in Figure 3a. As shown in Figure 3a, $J_{Ca,av}$ increased with an increase in SA concentration. This increase may be because the utilization efficiency of Ca^{2+} increases and the structure of CARP becomes looser with an increase in the SA concentration [3,7], resulting in an increase in the Ca^{2+} RSD. Moreover, as shown in the circular plots of Figure 3a, the ability of the CARP to attract Ca^{2+} ($W_{Ca,r}/W_{Ca}$) increases with the increasing SA concentration, owing to the more negatively charged alginate anions, which can bond to reverse osmotic Ca^{2+} on the FO membrane surface on the feed side. Furthermore, Figure 3b shows the change in the RSF of Ca^{2+} in the FS without considering Ca^{2+} in the CARP, J_{Ca} as the FO progressed; additionally, the RSF of Ca^{2+} decreased gradually for 1.0 g/L SA solution as the FS, while it initially increased and then decreased after 3.5 h for 2 and 3 g/L SA solution as the FS. It is speculated that, as the SA concentration increases, the reverse osmotic Ca^{2+} can react with more alginate ions in the FO process; therefore, the Ca-Alg colloid in the CARP accumulated on the FO membrane surface became larger and was easily eliminated by cross flow, thereby reducing the low resistance of Ca^{2+} reverse osmosis and high RSF. However, at high SA concentrations, the accumulation of CARP on the FO membrane was more significant; therefore, the resistance of Ca^{2+} reverse osmosis increased and the RSF decreased as the FO progressed. Notably, those results obtained were based on the concentration range with 1.0–3.0 g/L and, for a much higher SA solution concentration, the specific results should be further investigated in the future.

3.2. Calcium Ions in the Ca-Alg Recovery Production

The Ca^{2+} content in the CARP was evaluated. After FO was completed, the Ca^{2+} content in the FS before and after the elution of CARP on the FO membrane surface was measured. Later, the Ca^{2+} content in the CARP, $W_{Ca,r}$ was calculated using the difference in both Ca^{2+} concentrations. The Ca^{2+} content measured in the FS after the elution of CARP, W_{Ca} is the total Ca^{2+} content from the reverse osmotic solution containing Ca^{2+} in the CARP formed on the FO membrane. As shown in Table 1, $W_{Ca,r}$ and W_{Ca} decreased; however, the percentage content of Ca^{2+} in the CARP to total osmosis content of Ca^{2+} ($W_{Ca,r}/W_{Ca}$) increased with increasing DS concentrations for 1.0 g/L SA solution as the FS (Figure 2a), which enhanced the resistance of Ca^{2+} RSD by electrostatic interaction and decreased the salt concentration difference. Furthermore, as shown in Figure 3a, the $W_{Ca,r}/W_{Ca}$ up to 36.28% increased significantly with increasing SA concentrations for 2.0 M $CaCl_2$ solution as the DS. Therefore, these results indicated that the Ca^{2+} content in the CARP cannot be neglected while evaluating Ca^{2+} RSD in the Ca-Alg production process via FO with Ca^{2+} reverse diffusion. Furthermore, the Ca^{2+} content rate of total reverse osmosis to SA in the FS (W_{Ca}/W_{SA}), as shown in Table 1, indicates that, the lower the Ca^{2+} concentration in the DS and the SA concentration in the FS, the higher the content rate of Ca^{2+} in the concentrated recycled Ca-Alg in the FS. Therefore, optimal concentration conditions of the DS and FS were present to obtain a high water flux and availability of calcium ions while producing the target product Ca-Alg via FO with Ca^{2+} reverse diffusion.

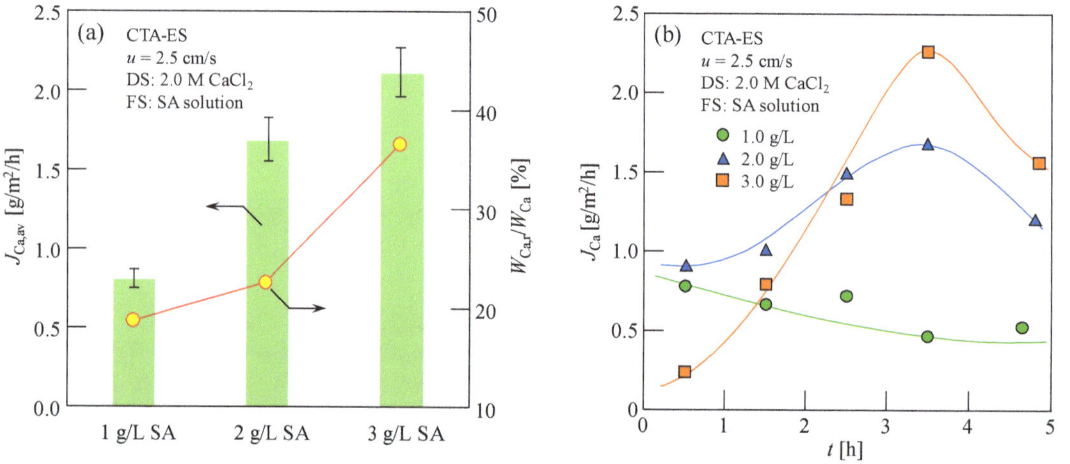

Figure 3. (**a**) Average RSF of Ca^{2+} ($J_{Ca,av}$) and percentage of Ca^{2+} content in the CARP formed on the FO membrane ($W_{Ca,r}$) to total Ca^{2+} content (W_{Ca}) in the FS containing Ca^{2+} in the CARP, $W_{Ca,r}/W_{Ca}$ for various SA solutions as the FS; (**b**) RSF of Ca^{2+} without considering Ca^{2+} in the CARP, J_{Ca}, as the FO progressed. Here, 1–3 g/L SA solution and 2.0 M $CaCl_2$ solution were used as the FS and DS, respectively; for the flow rates of 2.5 cm/s, u along opposite directions on both sides of the membrane was used; CTA-ES, cellulose triacetate with embedded polyester screen support.

The average water flux, $J_{w,av}$, in the FO is also shown in Table 1. The average water fluxes, $J_{w,av}$, are 6.63, 9.30, and 9.58 L/m^2/h in the FO of 1.0 g/L SA solution with 1.0, 2.0, and 3.0 M $CaCl_2$ DS, respectively, indicating that the water flux increased slightly with the increasing $CaCl_2$ concentration (>2.0 M). The trend of the water flux with the increasing $CaCl_2$ concentration differed from that of ultrapure water FO, in which the water flux increased dramatically with the increasing $CaCl_2$ concentration (1.0–3.0 M) from 10.66 to 19.29 L/m^2/h. Therefore, the higher the DS concentration, the faster the decrease in the increase in water flux in the FO of SA solution. However, $J_{w,av}$ for 1.0 g/L SA solution as the FS (9.30 L/m^2/h) was higher than those for 2.0 and 3.0 g/L SA (7.45 and 7.56 L/m^2/h, respectively), indicating that the water flux is not significantly influenced by the SA concentration. The water flux was dominated by the osmotic pressure difference between both sides of the FO membrane at the beginning of the FO. However, Ca-Alg CARP was formed on the FO membrane on the feed side owing to the interaction between SA and Ca^{2+} reverse osmosis as the FO progressed, thereby resulting in a significant transmembrane osmotic pressure difference. Additionally, the higher the DS concentration, the denser the membrane fouling of the gel layer formed on the FO membrane.

The Ca-Alg gel layer of the CARP formed on the FO membrane surface on the feed side was further analyzed using SEM and EDS. SEM morphology and EDS of Ca and Na elements in the CARP surface toward the feed side formed for 1.0 g/L SA as the FS and

1.0 and 3.0 M $CaCl_2$ as the DS to reveal the difference between low and high concentration draw solutes are shown in Figure 4. As shown in Figure 4(a1,b1), the CARP for 3 M DS is more loose than that for 1 M DS, along with a large Ca-Alg colloid formed for 3 M DS, suggesting that, the lower the concentration of DS, the stronger the molecular sieve role of the Ca-Alg formed on the FO membrane surface in the FS for resisting Ca^{2+} RSD. Therefore, for high DS concentrations, the electrostatic repulsion of CARP and the transmembrane salt osmotic pressure difference may dominate the Ca^{2+} RSD. In addition, as shown in Figure 4(a2,b2), the higher the DS concentration, the greater the Ca^{2+} content on the CARP surface, similar to the foregoing results, which may form large Ca-Alg colloids.

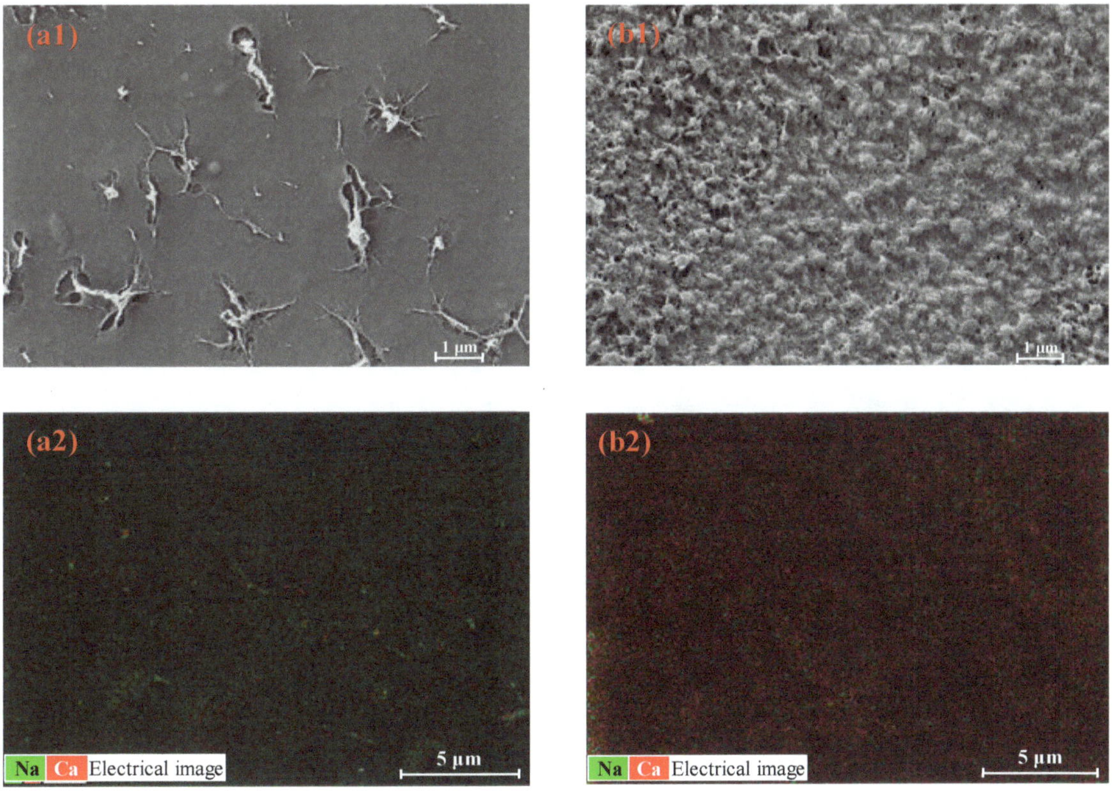

Figure 4. Scanning electron microscopy (SEM) morphology (**a1**,**b1**) and energy-dispersive spectroscopy (EDS) (**a2**,**b2**) of Ca and Na elements in the CARP surface towards the feed side. (**a1**,**a2**) represent the images for CARP formed for 1 M $CaCl_2$ as the draw solute, while (**b1**,**b2**) represent the images for CARP formed for 3 M $CaCl_2$ as the draw solute.

Furthermore, Figure 5 shows the SEM and EDS results of Ca and Na elements in the cross-sectional profile of CARP formed via FO with 1.0 g/L SA solution as the FS and 1.0 and 3.0 M $CaCl_2$ solution as the DS. As shown in Figure 5(a1,b1), the CARP was thicker for 1.0 M $CaCl_2$ than for 3.0 M $CaCl_2$, indicating that membrane fouling can be easily eliminated by cross flow for a high DS concentration. As shown in Figure 5(a2,b2), the previous results given in Figure 2a and Table 1, which showed that the Ca^{2+} content in the CARP decreases with the increasing DS concentration, were confirmed.

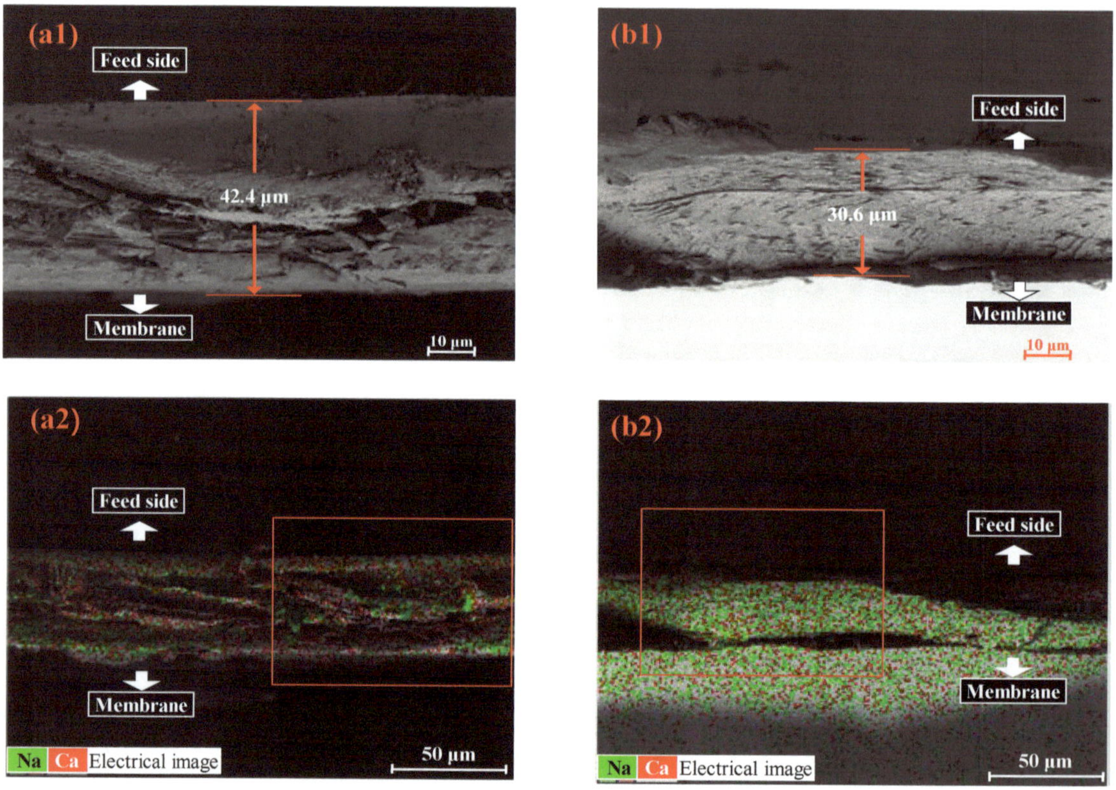

Figure 5. SEM morphology (**a1**,**b1**) and EDS images (**a2**,**b2**) of Ca and Na elements in the cross-sectional profile of CARP. (**a1**,**a2**) are CARP images for 1.0 g/L SA solution as the FS and 1 M $CaCl_2$ solution as the DS, where (**a1**) represents the red box in (**a2**); (**b1**,**b2**) are CARP images for 1.0 g/L SA solution as the FS and 3 M $CaCl_2$ solution as the DS, where (b1) represents the red box in (**b2**).

3.3. Ca-Alg Production Using ENM-TFC FO Membranes

The production properties of calcium alginate were also investigated for PSf or PVDF ENM-TFC FO membranes and compared to those obtained for commercial CTA-ES FO membranes, as shown in Table 2. To eliminate the effect of FS flow, a dead-end model (u = 3.0 cm/s for the DS side and u = 0 cm/s for FS side) was used [44]. The PSf ENM-TFC FO membranes were characterized by higher $J_{w,av}$ and $J_{ca,av}$ values than both the PVDF ENM-TFC FO and commercial CTA-ES FO membranes. $J_{w,av}$ and $J_{ca,av}$ values of PSf ENM-TFC FO membranes are higher than those of PVDF ENM-TFC FO membranes owing to the diluent concentration polarization of the support layer membrane caused by the small pore size of PVDF ENM (Figure 6). Therefore, considering the highly hydrophilic (low water contact angle) and elastic modulus (Table 3), PSf ENM is suggested to act as a support layer for FO membranes.

When the PSf ENM-TFC FO membrane was used with 1–3 M $CaCl_2$ as the DS and the 1 g/L SA as the FS, $J_{w,av}$ was proportional to the concentration of the DS, while $J_{ca,av}$ for 2 M $CaCl_2$ was higher than that for 3 M $CaCl_2$. This could be due to the resistance of the CARP Ca-Alg gel layer. In contrast, when the PSf ENM-TFC FO membrane was used with 2 M $CaCl_2$ as the DS and 1–3 g/L SA as the FS, there were little changes to both $J_{w,av}$ and $J_{ca,av}$, indicating that the SA concentration has only a slight effect on water flux and salt reverse diffusion. In conclusion, using TFC FO membranes with ENM as a support layer, the water flux and reverse solute flux can be increased, allowing control of the properties of the Ca-Alg production through the FO process.

Table 2. Average water flux, $J_{w,av}$, and reverse solute flux (RSF), $J_{ca,av}$, of Ca^{2+} in forward osmosis (FO) recovery of Ca-Alg after 4 h of FO with various FO membranes. Here, a 1–3 M $CaCl_2$ draw solution (DS), 1–3 g/L sodium alginate (SA) feed solution (FS), and the dead-end model (u = 3.0 cm/s for the DS side and u = 0 cm/s for the FS side) were used.

FO Membranes	Conc. of DS [M]	Conc. of SA [g/L]	$J_{w,av}$ [L/(m^2·h)]	$J_{ca,av}$ [g/(m^2·h)]
CTA-ES	2	1	4.22 ± 0.17	0.39 ± 0.06
PVDF ENM-TFC	2	1	3.98 ± 0.03	0.91 ± 0.09
	2	1	5.07 ± 0.14	1.13 ± 0.05
	1	1	4.19 ± 0.04	0.91 ± 0.32
PSf ENM-TFC	3	1	5.18 ± 0.38	0.63 ± 0.08
	2	2	5.94 ± 0.02	1.30 ± 0.22
	2	3	4.82 ± 0.11	1.11 ± 0.18

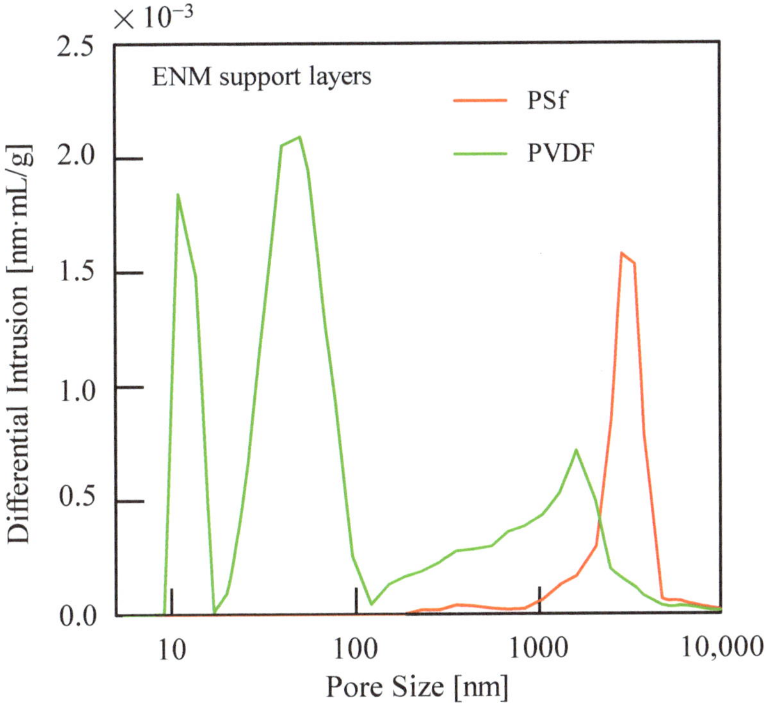

Figure 6. Pore sizes of electrospinning nanofiber membrane (ENM) support layers.

Table 3. Typical parameters of the ENM support layers fabricated in this study.

ENM Support Layer	Average Pore Size [μm]	Water Contact Angle [°]	Tensile Strength [MPa]	Elastic Modulus [MPa]
PSf	3.90 ± 0.47	42.90 ± 0.54	6.79 ± 0.59	160.51 ± 3.08
PVDF	0.51 ± 0.04	61.60 ± 2.08	6.58 ± 0.80	79.56 ± 2.31

3.4. Mechanism Analysis

Based on the above results, we have proposed four mechanisms on the effects of the CARP on the reverse diffusion of Ca^{2+} in the DS in Ca-Alg production via FO with alginate solution as the FS, as shown in Figure 7. The reverse osmotic Ca^{2+} in the DS through the FO membrane reacts with alginate in the FS and the CARP (Ca-Alg) is formed on the FO membrane on the feed side (Figure 7a), which influences the mass transfer of reverse osmotic Ca^{2+} from the DS. The molecular sieve role [45] of the CARP on the FO membrane surface in the feed side (Figure 7b) may play a major role in the low concentration of alginate in the FS. There is a dependence of electrostatic interaction [3,7,8] on the characteristics and number of charges in the gel layer of Ca-Alg CARP colloids (Figure 7c) with the formation of larger Ca-Alg CARP colloids on the FO membrane surface on the feed side because of the increase in alginate concentration and RSD of Ca^{2+}. Notably, the driving force of the salt concentration difference between the two sides of the FO membrane decreases because of the accumulation of reverse osmotic salt ions such as Ca^{2+} from the DS and the concentration of alginate ion (A^{n+}) on the membrane surface on the feed side (Figure 7d) [46,47]. Designing a novel FO membrane such as the ENM-TFC FO membrane is the most important strategy for Ca-Alg recovery production using an alginate solution FO process with Ca^{2+} reverse solute diffusion (Figure 7e) to regulate water and Ca^{2+} flux.

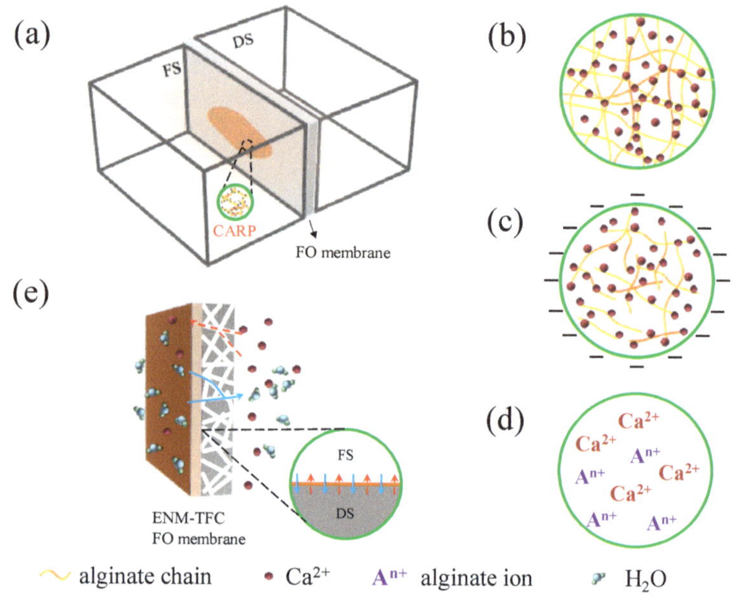

Figure 7. Schematic diagram of the mechanism analysis showing that Ca-Alg recovery production (CARP) affects the reverse diffusion of Ca^{2+} in the draw solution (DS) in calcium alginate production via forward osmosis (FO) with alginate solution as the feed solution (FS). (**a**) CARP on the FO membrane in FO cell; (**b**) image of CARP with molecular sieve role; (**c**) electrification of CARP colloids; (**d**) ions in CARP contributing to osmotic pressure; (**e**) visualization of a thin film composite FO membrane with an electrospinning nanofiber membrane support layer (ENM-TFC FO membrane) regulating water and salt flux.

4. Conclusions

In this study, a novel Ca-Alg production method was proposed, in which Ca-Alg in the FS was formed and concentrated via FO with Ca^{2+} reverse osmosis using SA solution as the FS and $CaCl_2$ solution as the DS. An abnormal RSD was found, with the average RSF decreasing with increasing $CaCl_2$ concentrations (1–3 M) for 1.0 g/L SA solution, while the RSD increased with increasing alginate concentrations (1.0–3.0 g/L) for 2.0 M $CaCl_2$ solution. These results were attributed to the formation of the CARP on the FO membrane surface on the feed side. The findings suggested that the Ca^{2+} content in the CARP must be considered to evaluate the actual Ca^{2+} RSF because the percentage of total Ca^{2+} reverse osmosis reached 36.28%. Moreover, optimal concentration conditions of the DS and FS are required to prepare the target product Ca-Alg via FO with Ca^{2+} reverse diffusion. SEM and EDS analyses also verified the existence of CARP and its Ca^{2+} content. In addition, the PSf ENM-TFC FO membrane was characterized by high water and reverse salt flux of Ca^{2+}. As such, PSf ENM is proposed as a support layer for TFC FO membranes because of its large pore size and its hydrophilic and mechanical properties. Four mechanisms were proposed to explain the effects of Ca-Alg production via the FO process with the reverse osmosis of Ca^{2+} from DS, such as the molecular sieve role, electrification of colloids, osmotic pressure of ions in CARP, and specific FO membrane structure. Future studies will focus on the alginate extracted from AGS to prepare and recycle Ca-Alg using the proposed method, other calcium salts as draw solutes, membrane fouling, as well as the evaluation of its contribution to the circular economy.

Author Contributions: Conceptualization, D.-Q.C.; methodology, D.-Q.C.; validation, D.-Q.C.; investigation, K.T., C.C., J.-L.H. and F.T.; data curation, D.-Q.C.; writing—original draft preparation, D.-Q.C.; writing—review and editing, D.-Q.C.; supervision, D.-Q.C.; project administration, D.-Q.C., W.-Y.Z. and X.-D.H. All authors have read and agreed to the published version of the manuscript.

Funding: This research was partially funded by the Beijing Nova Program (Z211100002121154), the Beijing Natural Science Foundation (8222043), the Project of Cultivation for young top-motch Talents of Beijing Municipal Institutions, the BUCEA Post Graduate Innovation Project, and Beijing Energy Conservation and Sustainable Urban and Rural Development Provincial and Ministry Co-Construction Collaboration Innovation Center.

Institutional Review Board Statement: Not applicable.

Data Availability Statement: Not applicable.

Conflicts of Interest: The authors declare that they have no known competing financial interests or personal relationships that could have appeared to influence the work reported in this paper.

Nomenclature

A_m	effective area of forward osmosis membrane	$[m^2]$
C_t	concentration of Ca^{2+} in feed solution at time t	$[kg/m^2/s]$
C_{t+t_0}	concentration of Ca^{2+} in feed solution at time $(t + t_0)$	$[kg/m^2/s]$
J_{Ca}	reverse solute flux of Ca^{2+} without considering Ca^{2+} in Ca-Alg recovery production passing through the forward osmosis membrane per unit membrane area per unit time at time t	$[kg/m^2/s]$
$J_{Ca,av}$	total average reverse solute flux of Ca^{2+} with considering Ca^{2+} in Ca-Alg recovery production passing through the forward osmosis membrane per unit membrane area from time $t = 0$ to t	$[kg/m^2/s]$
J_w	volume of water passing through the forward osmosis membrane per unit membrane area per unit time at time t	$[m/s]$
$J_{w,av}$	average volume of water passing through the forward osmosis membrane per unit membrane area from time $t = 0$ to t	$[m/s]$
t	forward osmosis time	[s]
t_0	time interval between two consecutive samplings	[s]
V_{DS}	volume of draw solution at time t	$[m^3]$
V_t	volume of feed solution at time t	$[m^3]$
V_{t+t_0}	volume of feed solution at time $(t + t_0)$	$[m^3]$
$W_{Ca,r}$	content of Ca^{2+} in Ca-Alg recovery production formed on the forward osmosis membrane	[kg]
W_{Ca}	total Ca^{2+} content of reverse osmosis containing Ca^{2+} in Ca-Alg recovery production formed on the forward osmosis membrane	[kg]
W_{SA}	content of SA in the feed solution	[kg]

References

1. Lee, K.Y.; Mooney, D.J. Alginate: Properties and biomedical applications. *Prog. Polym. Sci.* **2012**, *37*, 106–126. [CrossRef]
2. van Loosdrecht, M.C.M.; Brdjanovic, D. Anticipating the next century of wastewater treatment: Advances in activated sludge sewage treatment can improve its energy use and resource recovery. *Science* **2014**, *344*, 1452–1453.
3. Cao, D.Q.; Hao, X.D.; Wang, Z.; Song, X.; Iritani, E.; Katagiri, N. Membrane recovery of alginate in an aqueous solution by the addition of calcium ions: Analyses of resistance reduction and fouling mechanism. *J. Membr. Sci.* **2017**, *535*, 312–321.
4. van der Roest, H.; van Loosdrecht, M.C.M.; Langkamp, E.J.; Uijterlinde, C. Recovery and reuse of alginate from granular Nereda sludge. *Water* **2015**, *21*, 48.
5. Amorim de Carvalho, C.D.; Ferreira dos Santos, A.; Tavares Ferreira, T.J.; Sousa Aguiar Lira, V.N.; Mendes Barros, A.R.; Bezerra dos Santos, A. Resource recovery in aerobic granular sludge systems: Is it feasible or still a long way to go? *Chemosphere* **2021**, *274*, 129881.
6. Feng, C.J.; Lotti, T.; Canziani, R.; Lin, Y.M.; Tagliabue, C.; Malpei, F. Extracellular biopolymers recovered as raw biomaterials from waste granular sludge and potential applications: A critical review. *Sci. Total Environ.* **2021**, *753*, 142051.
7. Cao, D.Q.; Jin, J.Y.; Wang, Q.H.; Song, X.; Hao, X.D.; Iritani, E.; Katagiri, N. Ultrafiltration recovery of alginate: Membrane fouling mitigation by multivalent metal ions and properties of recycled materials. *Chinese J. Chem. Eng.* **2020**, *28*, 2881–2889.
8. Cao, D.Q.; Sun, X.Z.; Jin, J.Y.; Fang, X.M.; Yang, X.X.; Hao, X.D. Recovery of extracellular polymeric substance: Impact factors in forward osmosis separation of sodium alginate. *Environ. Eng.* **2020**, *38*, 1–6. (In Chinese)
9. Cao, D.Q.; Yang, X.X.; Jing, J.Y.; Sun, X.Z.; Hao, X.D. A Method for Alginate Concentration and Recovery via Forward Osmosis Using Reverse Diffusion of Draw Solute. C.N. Patent 202010189536.8, 18 March 2020. (In Chinese).
10. Cao, D.Q.; Sun, X.Z.; Zhang, W.Y.; Ji, Y.T.; Yang, X.X.; Hao, X.D. News on alginate recovery by forward osmosis: Reverse solute diffusion is useful. *Chemosphere* **2021**, *285*, 131483.
11. Ino, K.; Fukuda, M.T.; Hiramoto, K.; Taira, N.; Nashimoto, Y.J.; Shiku, H.S. Fabrication of three-dimensional calcium alginate hydrogels using sacrificial templates of sugar. *J. Biosci. Bioeng.* **2020**, *130*, 539–544. [CrossRef]
12. Varaprasad, K.; Jayaramudu, T.; Kanikireddy, V.; Toro, C.; Sadiku, E.R. Alginate-based composite materials for wound dressing application: A mini review. *Carbohyd. Polym.* **2020**, *236*, 116025.
13. Cai, J.; Chen, X.J.; Wang, X.J.; Tan, Y.L.; Ye, D.D.; Jia, Y.T.; Liu, P.F.; Yu, H. High-water-absorbing calcium alginate fibrous scaffold fabricated by microfluidic spinning for use in chronic wound dressings. *RSC ADV.* **2018**, *8*, 39463–39469.
14. Kashima, K.; Inage, T.; Yamaguchi, Y.; Imai, M. Tailorable regulation of mass transfer channel in environmentally friendly calcium alginate membrane for dye removal. *J. Environ. Chem. Eng.* **2021**, *9*, 105210. [CrossRef]
15. Zhou, Q.Z.; Wang, W.Z.; Liu, F.M.; Chen, R. Removal of difenoconazole and nitenpyram by composite calcium alginate beads during apple juice clarification. *Chemosphere* **2022**, *286*, 131813. [CrossRef]
16. Duc, T.H.; Vu, T.K.; Dang, C.T.; Nguyen, V.H.; La, D.D.; Kim, G.M.; Chang, S.W.; Bui, X.T.; Dang, T.D.; Nguyen, D.D. Synthesis and application of hydrogel calcium alginate microparticles as a biomaterial to remove heavy metals from aqueous media. *Environ. Technol. Inno.* **2021**, *22*, 101400. [CrossRef]
17. Song, D.; Park, S.J.; Kang, H.W.; Park, S.B.; Han, J.I. Recovery of lithium(i), strontium(ii), and lanthanum(iii) using ca–alginate beads. *J. Chem. Eng. Data.* **2013**, *58*, 2455–2464. [CrossRef]
18. Khajavi, P.; Keshtkar, A.R.; Moosavian, M.A. The optimization of U(VI) removal by a novel amidoximated modified calcium alginate gel bead with entrapped functionalized SiO_2 nanoparticles. *Prog. Nucl. Energ.* **2021**, *140*, 103887.
19. Isik, Z.; Saleh, M.; Dizge, N. Adsorption studies of ammonia and phosphate ions onto calcium alginate beads. *Surf. Interfaces* **2021**, *26*, 101330.
20. Sun, C.; Zhan, H.H.; Wang, L.J. Enhanced aerobic sludge granulation by seeding concentrated activated sludge with ca-alginate gel. *Water Environ. Res.* **2017**, *89*, 2078–2087. [CrossRef]
21. Meng, F.; Li, M.; Wang, H.; Xin, L.; Wu, X.; Liu, X. Encapsulating microscale zero valent iron-activated carbon into porous calcium alginate for the improvement on the nitrate removal rate and Fe^0 utilization factor. *Micropor. Mesopor. Mat.* **2020**, *307*, 110522.
22. Su, C.; Sun, X.J.; Mu, Y.Z.; Li, P.W.; Li, J.; Fan, P.S.; Zhang, M.X.; Wang, M.Y.; Chen, X.G.; Feng, C. Multilayer calcium alginate beads containing Diatom Biosilica and *Bacillus subtilis* as microecologics for sewage treatment. *Carbohyd. Polym.* **2021**, *256*, 117603.
23. Zhang, Y.; Zhao, K.Y.; Zhang, Z.J.; Xie, H.K.; Li, Z.Y.; Lin, Z.; Mo, C.; Yang, W.D.; Wang, X.L.; Wei, J.F. Polypropylene non-woven supported calcium alginate hydrogel filtration membrane for efficient separation of dye/salt at low salt concentration. *Desalination* **2020**, *500*, 114845.
24. Zhang, Y.; Zhao, K.Y.; Yang, Z.H.; Zhang, Z.J.; Guo, Z.L.; Chu, R.R.; Zhang, W.; Shi, W.X.; Li, J.C.; Li, Z.Y.; et al. Calcium alginate and barium alginate hydrogel filtration membrane coated on fibers for molecule/ion separation. *Separ. Purif. Technol.* **2021**, *270*, 118761.
25. Xu, S.; Tabakovic, A.; Liu, X.; Schlangen, E. Calcium alginate capsules encapsulating rejuvenator as healing system for asphalt mastic. *Constr. Build. Mater.* **2018**, *169*, 379–387.
26. Bao, S.W.; Liu, Q.T.; Rao, W.H.; Yu, X.; Zhang, L. Synthesis and characterization of calcium alginate-attapulgite composite capsules for long term asphalt self-healing. *Constr. Build. Mater.* **2020**, *265*, 120779.
27. Rao, W.; Liu, Q.; Yu, X.; Wan, P.; Wang, H.; Song, J.; Ye, Q. Efficient preparation and characterization of calcium alginate-attapulgite composite capsules for asphalt self-healing. *Constr. Build. Mater.* **2021**, *299*, 123931. [CrossRef]

28. Pfaff, N.M.; Kleijn, J.M.; van Loosdrecht, M.C.M.; Kemperman, A.J.B. Formation and ripening of alginate-like exopolymer gel layers during and after membrane filtration. *Water Res.* **2021**, *195*, 116959. [CrossRef]
29. Kahrizi, M.; Lin, J.Y.; Ji, G.Z.; Kong, L.X.; Song, C.W.; Dumée, L.F.; Sahebi, S.; Zhao, S.F. Relating forward water and reverse salt fluxes to membrane porosity and tortuosity in forward osmosis: CFD modelling. *Separ. Purif. Technol.* **2020**, *241*, 11672. [CrossRef]
30. Zou, S.Q.; Qin, M.H.; He, Z. Tackle reverse solute flux in forward osmosis towards sustainable water recovery: Reduction and perspectives. *Water Res.* **2019**, *149*, 362–374. [CrossRef]
31. Chiao, Y.H.; Sengupta, A.; Ang, M.B.M.Y.; Chen, S.T.; Haan, T.Y.; Almodovar, J.; Hung, W.S.; Wickramasinghe, S.R. Application of zwitterions in forward osmosis: A short review. *Polymers* **2021**, *13*, 583.
32. Wu, Z.Y.; Zou, S.Q.; Zhang, B.; Wang, L.J.; He, Z. Forward osmosis promoted in-situ formation of struvite with simultaneous water recovery from digested swine wastewater. *Chem. Eng. J.* **2018**, *342*, 274–280.
33. Wu, S.M.; Zou, S.Q.; Yang, Y.L.; Qian, G.R.; He, Z. Enhancing the performance of an osmotic microbial fuel cell through self-buffering with reverse-fluxed sodium bicarbonate. *Chem. Eng. J.* **2018**, *349*, 241–248. [CrossRef]
34. Volpin, F.; Yu, H.; Cho, J.; Lee, C.; Phuntsho, S.; Ghaffour, N.; Vrouwenvelder, J.S.; Shon, H.K. Human urine as a forward osmosis draw solution for the application of microalgae dewatering. *J. Hazard. Mater.* **2019**, *378*, 120724. [CrossRef]
35. Park, H.B.; Kamcev, J.; Robeson, L.M.; Elimelech, M.; Freeman, B.D. Maximizing the right staff: The trade-off between membrane permeability and selectivity. *Science* **2017**, *356*, eaab0530.
36. Cao, D.Q.; Liu, X.D.; Han, J.L.; Zhang, W.Y.; Hao, X.D.; Iritani, E.; Katagiri, N. Recovery of extracellular polymeric substances from excess sludge using high-flux electrospun nanofiber membranes. *Membranes* **2023**, *13*, 74. [CrossRef]
37. Bui, N.; Lind, M.; Hoek, E.; McCutcheon, J. Electrospun nanofiber supported thin film composite membrane for engineered osmosis. *J. Membrane. Sci.* **2011**, *385*, 10–19.
38. Song, X.; Liu, Z.; Sun, D.D. Nano gives the answer: Breaking the bottleneck of internal concentration polarization with a nanofiber composite forward osmosis membrane for a high water production rate. *Adv. Mater.* **2011**, *23*, 3256–3260.
39. Pugua, J.; Kim, H.S.; Lee, K.J.; Kim, H. Low internal concentration polarization in forward osmosis membranes with hydrophilic crosslinked PVA nanofibers as porous support layer. *Desalination* **2014**, *336*, 24–31.
40. Ndiaye, I.; Vaudreuil, S.; Bounahmidi, T. Forward osmosis process: State-of-the-art of membranes. *Sep. Purif. Rev.* **2021**, *50*, 53–73. [CrossRef]
41. Tian, M.; Qiu, C.; Liao, Y.; Chou, S.; Wang, R. Preparation of polyamide thin film composite forward osmosis membranes using electrospun polyvinylidene fluoride (PVDF) nanofibers as substrates. *Sep. Purif. Technol.* **2013**, *118*, 727–736. [CrossRef]
42. Wang, W.; Guo, Y.; Liu, M.; Song, X.; Duan, J. Porous nano-hydroxyapatites doped into substrate for thin film composite forward osmosis membrane to show high performance. *Korean J. Chem. Eng.* **2020**, *37*, 1573–1584. [CrossRef]
43. Cao, D.Q.; Yang, X.X.; Yang, W.Y.; Wang, Q.H.; Hao, X.D. Separation of trace pharmaceuticals individually and in combination via forward osmosis. *Sci. Total Environ.* **2020**, *718*, 137366. [CrossRef]
44. Cao, D.Q.; Hao, X.D.; Wang, Q.H.; Fang, X.M.; Song, X. A Novel Setup and Method of Forward Osmosis Concentration. C.N. Patent 201910619042.6, 10 July 2019. (In Chinese).
45. Huang, J. Molecular sieving effect of a novel hyper-cross-linked resin. *Chem. Eng. J.* **2010**, *165*, 265–272. [CrossRef]
46. Nguyen, T.T.; Adha, R.S.; Lee, C.; Kim, D.H.; Kim, I.S. Quantifying the influence of divalent cations mass transport on critical flux and organic fouling mechanism of forward osmosis membrane. *Desalination* **2021**, *512*, 115146.
47. Song, L.; Heiranian, M.; Elimelech, M. True driving force and characteristics of water transport in osmotic membranes. *Desalination* **2021**, *520*, 115360. [CrossRef]

membranes

MDPI

Article

Separation Properties of Plasmid DNA Using a Two-Stage Particle Adsorption-Microfiltration Process

Nobuyuki Katagiri [1,*], Daisuke Shimokawa [2], Takayuki Suzuki [2], Masahito Kousai [2] and Eiji Iritani [2]

[1] Department of Environmental Technology, Meijo University, 1-501 Shiogamaguchi, Tempaku-ku, Nagoya 468-8502, Japan

[2] Department of Chemical Engineering, Nagoya University, Furo-cho, Chikusa-ku, Nagoya 464-8603, Japan

* Correspondence: katagiri@meijo-u.ac.jp; Tel.: +81-52-838-2368

Abstract: Plasmid DNA is used as a vector for gene therapy and DNA vaccination; therefore, the establishment of a mass production method is required. Membrane filtration is widely employed as a separation method suitable for the mass production of plasmid DNA. Furthermore, the separation of plasmid DNA using microfiltration and ultrafiltration membranes is being investigated. Because plasmid DNA has a circular structure, it undergoes significant deformation during filtration and easily permeates the membrane, hindering the selection of separation membranes based on molecular weight. In this study, we applied affinity microfiltration to plasmid DNA purification. α-Fe_2O_3 with an isoelectric point of approximately 8 and a particle size of 0.5 μm was selected as the ligand for two-stage affinity microfiltration of plasmid DNA. In the first stage of microfiltration, the experiment was conducted at a pH of 5, and a cake of α-Fe_2O_3 with bound plasmid DNA was obtained. Next, liquid permeation (pH 9 and 10) through the cake was performed to elute the bound plasmid DNA. Plasmid DNA was eluted during the early phase of liquid permeation at pH 10. Furthermore, agarose gel analysis confirmed the usefulness of the two-stage affinity microfiltration method with adsorption and desorption for plasmid DNA purification.

Keywords: microfiltration; adsorption; desorption; plasmid DNA; affinity; ligand

Citation: Katagiri, N.; Shimokawa, D.; Suzuki, T.; Kousai, M.; Iritani, E. Separation Properties of Plasmid DNA Using a Two-Stage Particle Adsorption-Microfiltration Process. *Membranes* **2023**, *13*, 168. https://doi.org/10.3390/membranes13020168

Academic Editors: Peter Czermak and Scott M. Husson

Received: 28 December 2022
Revised: 27 January 2023
Accepted: 27 January 2023
Published: 29 January 2023

1. Introduction

Recently, gene therapy and DNA vaccines have been actively developed for the treatment of various diseases [1]. Gene therapy requires a vector that acts as a carrier for gene replacement, and plasmids are used as non-viral vectors [2,3]. Plasmid DNA is an extranuclear gene that exists in bacteria such as *Escherichia coli* and replicates independently of chromosomal DNA. To use plasmid DNA in gene therapy, it is necessary to mass-produce pharmaceutical-grade plasmid DNA and develop industrial-scale isolation and purification methods [4,5]. Plasmid DNA purification starts with the process of extracting plasmid DNA from the inside of the bacterial cells, followed by lysis through the addition of chemicals, separation of the bacterial mass, and chromatography [6,7]. The use of hazardous substances that affect the human body is preferably avoided, and a safe, scalable, and cost-effective purification process for plasmid DNA needs to be developed.

Membrane processes have immense potential for large-scale plasmid purification. Several studies [8–15] have demonstrated that membrane-based processes are effective for the purification of plasmid DNA. Microfiltration membranes are mainly used to remove contaminants such as chromosomal DNA, proteins, and aggregates of bacteria because capturing nanosized plasmid DNA is difficult [11,14]. In contrast, ultrafiltration membranes that can capture nanosized particles are used to capture and purify or concentrate plasmid DNA [14]. However, as plasmid DNA has a circular structure and is significantly deformed during filtration, it may permeate the membrane, depending on the filtration conditions, even when an ultrafiltration membrane is used. Therefore, although the selection of separation membranes based on molecular weight is difficult, several studies have been conducted on the physical mechanisms governing DNA transmission and the effects of membrane pore size and operating conditions on the DNA sieving coefficient [9,10,12].

Affinity membrane filtration, in which a large ligand is used to selectively bind the desired materials in solution and is retained by a semipermeable membrane, is a promising purification technique for biopolymers [16,17]. Using this method, plasmid DNA can be adsorbed onto submicron-sized ligands and captured using microfiltration membranes. Since the performance of affinity filtration is significantly influenced by the specific binding interactions between the targeted material and the ligand, various types of ligands have been applied to attain a higher level of separation efficiency for biopolymers. Affinity substances for plasmid DNA include metal ions, metal oxides, peptides, and proteins [18–21]. In general, these ligands are used as a fixed layer; however, in this study, we investigated a membrane filtration method in which metal oxide particles of a size that can be captured using a microfiltration membrane are added as ligands to a solution containing plasmid DNA.

E. coli is often used for the production of plasmid DNA, and several extraction methods have been investigated [22,23]. Cell disruption for plasmid DNA extraction should be performed to minimize damage to plasmid DNA and genomic DNA. Alkaline lysis is the most commonly used method for cell disruption; however, it has known limitations, including low plasmid DNA recovery and a time-consuming process. Haberl et al. showed that electroextraction is a swifter alternative to alkaline lysis for the extraction of plasmid DNA [22]. Padilla-Zamudio et al. showed that cell disruption in a bead mill was more efficient in releasing plasmid DNA than alkaline lysis [23]. High pressure is also effective for cell disruption, and it is known that metabolites such as nucleic acids can be extracted from *E. coli* cells at pressures above 600 kPa [24]. Each extraction method has advantages and disadvantages; therefore, to establish a highly efficient purification method for plasmid DNA, an examination of the extraction of plasmid DNA, including its separation properties after cell disruption, is necessary.

In the present study, the application of affinity microfiltration to plasmid DNA purification and the search for ligands was examined. In addition to the selectivity behaviors in the binding process of plasmid DNA to the ligand and the elution process of the bound plasmid DNA, the membrane filtration behaviors of plasmid and ligand were also investigated in this system. Furthermore, we investigated the cell disruption method for plasmid DNA extraction and the membrane filtration properties of the disrupted suspension. The results of this study demonstrated the effectiveness of a two-stage microfiltration process, in which both the adsorption and desorption of plasmid DNA to large ligands exhibit immense potential for plasmid DNA purification.

2. Materials and Methods

2.1. Materials

A 3.0 kb plasmid DNA pBluescript II SK(+) was obtained from Stratagene Corp., San Diego, CA, USA. *Escherichia coli* DH5α (Nippon Gene Co. Ltd., Tokyo, Japan) was used as the host for the plasmid and grown at 303 K on an LB medium supplemented with the ampicillin antibiotic. The test solution was prepared by the following three steps: alkaline lysis of *E. coli* containing plasmid DNA, the addition of $CaCl_2$ for the removal of high molecular weight RNA [25], and the addition of ethanol for the concentration of nucleic acid. The plasmid DNA-containing sediment was dissolved in 10 mM Tris-HCl buffer (pH 5), and this solution, free of impurities such as proteins, was used for a two-stage affinity microfiltration experiment. The ligand employed in the experiments was α-Fe_2O_3 (particle size: 0.5 μm) provided by the Kojundo Chemical Lab. Co. Ltd., Saitama, Japan. A microelectrophoresis Mark II apparatus (Rank Brothers Ltd., Cambridge, UK) was used to determine the zeta potential of α-Fe_2O_3 particles.

2.2. Adsorption and Desorption Experiments

Plasmid DNA solutions of different concentrations were prepared and added to α-Fe_2O_3 slurries with known concentrations (0.1–80 mg/mL) to measure the adsorption properties. The solvents used for the solutions and slurries were pH 5–7 10 mM Tris-HCl buffer. An amount of 1 mL of each solution was maintained at a constant temperature of 298 K for 1 h, which was confirmed to be sufficient to achieve a quasi-steady state in the preliminary test. The desorption of the plasmid DNA adsorbed onto the particles was performed by changing the pH of the solution and allowing it to stand for 1 h. The amounts of adsorbed or desorbed plasmid DNA were determined from the concentrations of plasmid DNA in the solutions before and after the experiments using a spectrophotometer (UV-1800, Shimadzu Corp., Kyoto, Japan). The plasmid DNA used in the adsorption/desorption experiments was purified using a Qiagen plasmid midi kit.

2.3. Two-Stage Affinity Microfiltration Experiments

An unstirred batch filtration cell with an effective membrane area of 19.6 cm^2 was utilized in this study. Microfiltration experiments were performed in the dead-end filtration mode under constant pressure by applying compressed nitrogen gas [26]. The filtrate was collected in a reservoir placed on an electronic balance (Shimadzu Corp., Kyoto, Japan) connected to a personal computer to collect and record mass versus time data. The weights were converted to volumes using density correlations. A mixed cellulose ester microfiltration membrane (Advantec Toyo Co. Ltd., Tokyo, Japan) with a nominal pore size of 0.1 μm was employed in the experiments. In the first stage of microfiltration, the experiment was conducted under the condition of pH 5 using test solutions containing ligands (160 mL, the mass fraction of the solids s = 0.016), and a cake of α-Fe_2O_3 with plasmid bound was obtained. In the second stage of microfiltration, 10 mL of 1 M Tris-HCl buffer (pH 9) was added to the top of the cake and allowed to permeate after standing for 1 h. Next, the permeation of 2 M Tris-HCl buffer (pH 10, 50 mL) through the cake was performed to elute the bound plasmid DNA. The concentration of plasmid DNA in the permeate was measured at 260 nm using a spectrophotometer (UV-1800, Shimadzu Corp., Kyoto, Japan). The quality of the plasmid DNA was confirmed using agarose gel electrophoresis. A permeate of 5 μL was mixed with 1 μL of 6×loading buffer (0.25% bromophenol blue, 0.25% xylenecyanol, and 5 mM EDTA in 30% glycerol) and subjected to electrophoresis. Electrophoresis was performed in 0.6% (*w*/*v*) agarose (Nippon Gene Co. Ltd., Tokyo, Japan) containing ethidium bromide for 1 h at 100 V for using a submarine electrophoresis system (Nihon Eido Co.Ltd., Tokyo, Japan). Gels were placed on a UV table (Atto Corp., Tokyo, Japan), and photographs were taken with Polaroid (Funakoshi Co. Ltd., Tokyo, Japan). OneSTEP Marker 1 (λ/Hind III digest, Nippon Gene Co. Ltd., Tokyo, Japan) was used as a molecular marker. The two-stage affinity microfiltration experiments were performed more than three times to ensure the reproducibility of the results.

2.4. Cell Disruption Experiments

Three mechanical cell disruption methods were investigated to extract plasmid DNA from *E. coli* cells. The *E. coli* cells, after cultivation, were collected using centrifugation (3000 rpm, 15 min) and suspended in pure water to prepare a suspension (3.0×10^8 cell/mL). 10 mL of *E. coli* suspension was disrupted with an ultrasonic homogenizer (UP-200S, Dr. Hielscher GmbH, Stuttgart, Germany) at an operating frequency of 24 kHz and a nominal load power output of 200 W for 60 s. Cell disruption using a bead mill was performed by setting a 2 mL tube (zirconia beads of 1 mm diameter, 60 beads, Sarstedt Inc., Newton, MA, USA) in a Delta Mixer (Se-08, Taitec Corp., Tokyo, Japan) and shaking for 30 min at 3000 rpm. An attempt was made to extract the plasmid DNA from the cells using electroporation. The condition was as follows: 0.2 cm-gap sterile electroporation cuvette, pulse number 10, voltage 500 V, pulse length 100 ms, and interval 0.1 s using the Gene Pulser Xcell Electroporation System (Bio-Rad Laboratories, Inc., Hercules, CA, USA). Photomicrographs of *E. coli* after disruption were obtained using a digital photomicroscope (BA210EINT,

Shimadzu Rika Corp., Tokyo, Japan). A suspension of disrupted *E. coli* cells was subjected to affinity filtration after removing solids with constant pressure microfiltration (p = 49 kPa, mixed cellulose ester membrane with 0.1 μm).

3. Results and Discussion

3.1. Adsorption and Desorption Properties of Plasmid DNA

Figure 1 shows the pH dependence of the zeta potential of α-Fe_2O_3 particles used as ligands. The isoelectric point is approximately pH 8, and it is positively charged at a pH lower than eight and negatively charged at a pH above eight. Since plasmid DNA is a polyanion, it is expected to be adsorbed on the surface of positively charged α-Fe_2O_3 by setting the pH below seven. In contrast, in a solution environment with a pH greater than nine, an electrostatic repulsive force acts between α-Fe_2O_3 and plasmid DNA.

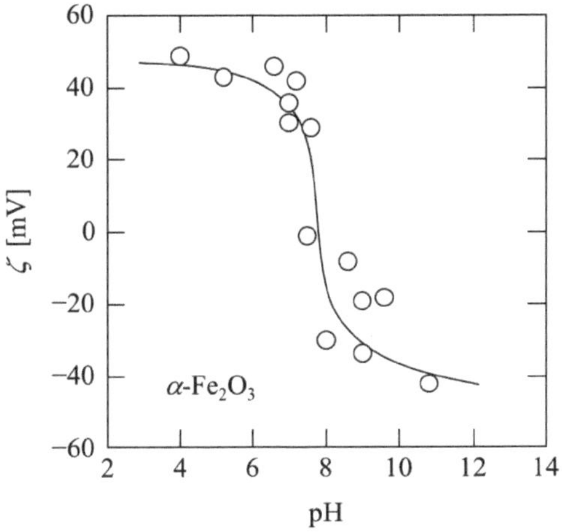

Figure 1. pH dependence of zeta potential of α-Fe_2O_3 particles.

The adsorption isotherms of plasmid DNA onto α-Fe_2O_3 were obtained through batch adsorption experiments, and the results at pH 7 are shown in Figure 2. The experimental data were approximated using the Langmuir adsorption isotherm equation, represented by

$$W = \frac{aW_sC}{1+aC}, \quad (1)$$

where W is the amount of plasmid DNA adsorbed, a is the Langmuir adsorption constant, W_s is the maximum adsorption capacity of α-Fe_2O_3 for plasmid DNA, and C is the equilibrium concentration of plasmid DNA in the solution. The solid line in the figure represents the calculated value based on Equation (1). This result is consistent with the findings of Liu et al., in which the DNA adsorption behavior of modified magnetic nanoparticles follows the Langmuir isotherm model [27]. As can be seen from the figure, the amount of adsorption is large, even at extremely low concentrations, and the affinity of plasmid DNA for α-Fe_2O_3 is extremely high. However, the maximum adsorption amount of plasmid DNA on the modified magnetic nanoparticles used by Liu et al. was approximately 10 times larger than that on the iron oxide particles we used. It is expected that the adsorption amount of plasmid DNA can be increased by modifying the surface of the iron oxide particles.

In Figure 3, the maximum adsorption capacity W_s is plotted against the pH of the solution. The amount of plasmid DNA adsorbed is strongly dependent on pH and decreases with increasing pH at pH 5–7. The W_s at pH 5 was approximately twice that at pH 7. By lowering the pH, more plasmid DNA can be adsorbed; however, if it is extremely low, plasmid DNA may deteriorate. Therefore, the first stage of microfiltration involving the binding of plasmid DNA was performed at pH 5.

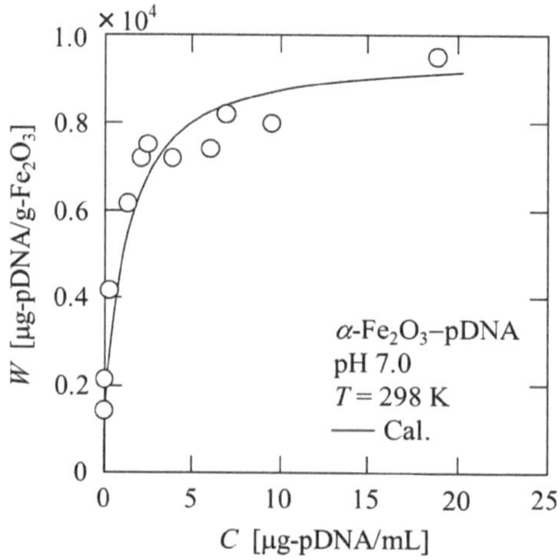

Figure 2. Property of adsorption of plasmid DNA onto α-Fe_2O_3.

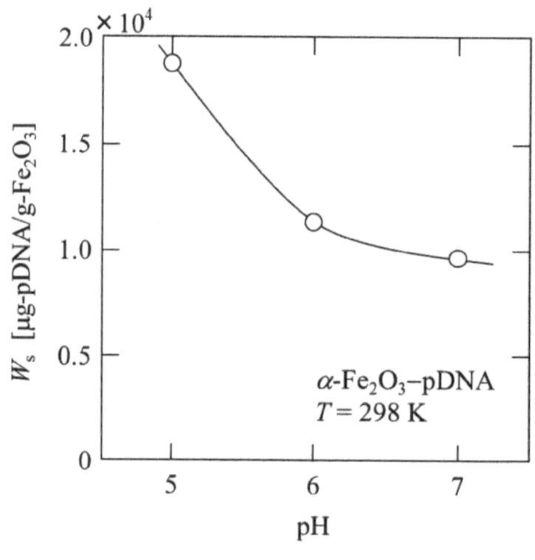

Figure 3. Dependence of adsorption capacity of plasmid DNA on the pH of the solution.

Figure 4 shows the effects of pH on the desorption of plasmid DNA from α-Fe_2O_3. After the plasmid was adsorbed onto α-Fe_2O_3, the pH of the solution was gradually increased, and the desorption efficiency D of the plasmid DNA was determined by measuring the amount of desorbed plasmid DNA that migrated into the solution. The plasmid DNA was desorbed at a pH above the isoelectric point of α-Fe_2O_3, and the desorption efficiency D was approximately 100% above pH 10. At pH 10, the plasmid DNA was desorbed from α-Fe_2O_3 particles (0.125 mg/mL) and recovered as a solution with a concentration of 1.6 μg/mL. Impurities can be separated using adsorption filtration of the plasmid DNA, and subsequently, the plasmid DNA can be recovered using desorption filtration. Therefore, α-Fe_2O_3 particles are determined to be suitable as a ligand. The second stage of microfiltration involving the desorption of plasmid DNA was performed using liquid permeation with a stepwise increase in pH.

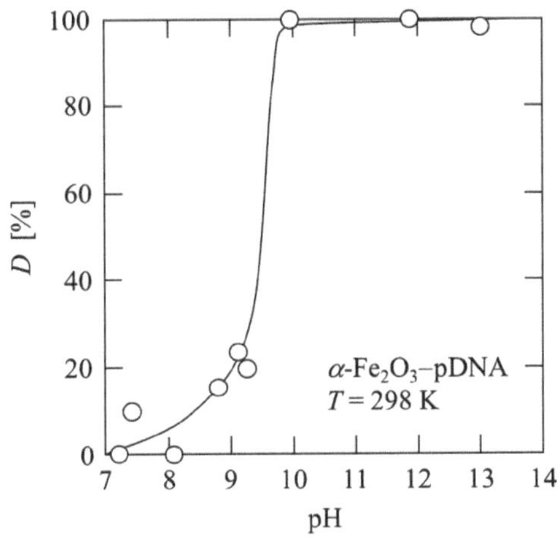

Figure 4. Desorption property of plasmid DNA corresponding to the pH change of the solution.

3.2. Two-Stage Affinity Microfiltration Properties of Plasmid DNA

Although plasmid DNA permeates the microfiltration membrane, the ligand α-Fe_2O_3 particles are captured using the 0.1 µm microfiltration membrane, forming a filter cake. Plasmid DNA can be purified using two-stage microfiltration by adsorbing the plasmid DNA onto the α-Fe_2O_3 cake to separate impurities and subsequently desorbing from the α-Fe_2O_3 cake. Based on the results of the adsorption experiments, the amount of α-Fe_2O_3 required to adsorb almost 100% of the plasmid DNA in the test solution prepared from *E. coli* was calculated, and adsorption microfiltration experiments were performed. Typical data of the microfiltration experiments of α-Fe_2O_3 and α-Fe_2O_3 with plasmid DNA-bound slurries at pH 5 are plotted in Figure 5 in the form of the reciprocal filtration rate $(d\theta/dv)$ against the filtrate volume v per unit effective membrane area. For the filtration of the α-Fe_2O_3 slurry, the plots appeared to be linear according to the Ruth filtration rate equation, expressed as [28]

$$\frac{d\theta}{dv} = \frac{\mu\rho s\alpha_{av}}{p(1-ms)}(v+v_m), \tag{2}$$

where θ is the filtration time, μ is the viscosity of the filtrate, ρ is the density of the filtrate, s is the mass fraction of the solids in the slurry, p is the applied filtration pressure, v_m is the fictitious filtrate volume per unit effective membrane area, and m is the ratio of the mass of the wet cake to the mass of the dry cake. The average specific cake resistance α_{av} was calculated from Equation (2) using the slope of the plot. In contrast, for the filtration of α-Fe_2O_3 with the plasmid DNA-bound slurry, cake formation was significantly affected by particle settling. After the formation of the filter cake, the supernatant fluid permeated the filter cake. During this period, $d\theta/dv$ remained approximately constant. From this constant value $(d\theta/dv)_p$, α_{av} can be calculated as

$$\alpha_{av} = \frac{p}{\mu w}\left\{\left(\frac{d\theta}{dv}\right)_p - \left(\frac{d\theta}{dv}\right)_m\right\}, \tag{3}$$

where w is the net solid mass of the entire cake per unit effective membrane area, and $(d\theta/dv)_m$ is the reciprocal filtration rate, which is equivalent to the flow resistance of the membrane. It was observed that average specific cake resistance decreased by approximately 1/3 from 3.9×10^{12} m/kg to 1.0×10^{12} m/kg because of the binding of plasmid DNA to α-Fe_2O_3. This phenomenon is attributed to the charge neutralization of positively charged α-Fe_2O_3 by the polyanion plasmid DNA, resulting in floc formation and coarsening. In addition, the filtrate did not contain plasmid DNA, as shown in Figure 6, lane 7.

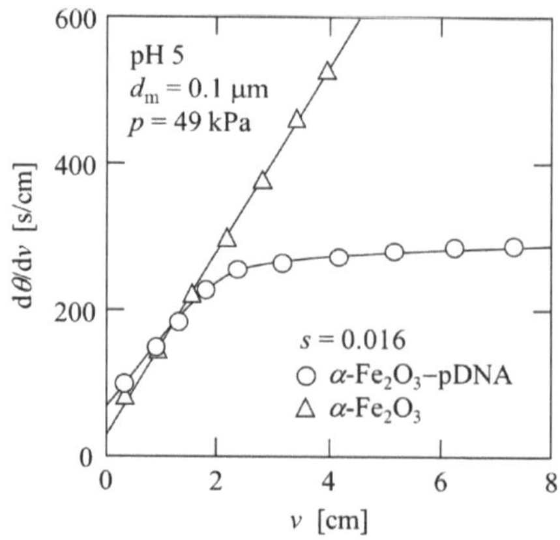

Figure 5. Relation between reciprocal filtration rate and filtrate volume per unit membrane area in adsorption-filtration.

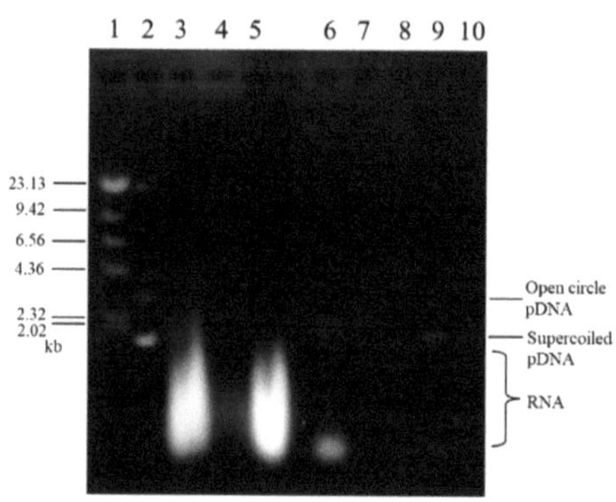

Figure 6. Agarose gel analysis: lane 1, DNA molecular weight standard; lane 2, plasmid DNA; lane 3, *E. coli* lysate; lane 4, supernatant ($CaCl_2$ addition); lane 5, sediment ($CaCl_2$ addition); lane 6, sediment (ethanol addition); lane 7, filtrate (pH 5); lane 8, permeate (pH 9); lane 9, permeate (pH 10 early phase); lane 10, permeate (pH 10 later phase).

In Figure 7, $d\theta/dv$ and the optical density at a wavelength of 260 nm (OD_{260}) of the permeate are plotted against the permeate volume v per unit effective membrane area in the elution process. Upon changing the pH of the permeate from 9 to 10, the value of $d\theta/dv$ changed from 400 s/m to 600 s/cm, indicating a change in the cake structure. A higher pH resulted in a higher average specific cake resistance α_{av}, as determined from Equation (3). Furthermore, the variation in OD_{260} of the permeate showed that plasmid DNA was eluted in the early phase of the pH 10 liquid permeation (2 M Tris-HCl buffer). This was confirmed using agarose gel analysis, as shown in Figure 6 (lanes 9 and 10). The test solution with OD_{260} = 4.05 was recovered as a solution with OD_{260} = 1.05 using a two-stage affinity microfiltration.

Figure 6 shows the results of the agarose gel electrophoresis of the solutions obtained after each treatment. The solution (lane 6) that was subjected to adsorption filtration after alkaline lysis of *E. coli* (lane 3) and the addition of $CaCl_2$ (lanes 4 and 5) were found to contain plasmid DNA and a large amount of low-molecular-weight RNA. Neither plasmid DNA nor RNA was confirmed in the filtrate (lane 7) of the adsorption filtration; therefore, both nucleic acids are assumed to be adsorbed by α-Fe_2O_3 and exist in the cake on the membrane surface. In the desorption filtration of plasmid DNA, a small amount of low-molecular-weight RNA was confirmed in the permeate of pH 9 (lane 8). Subsequently, plasmid DNA and RNA were confirmed in the initial permeate at pH 10 (lane 9), and only plasmid DNA was confirmed in the subsequent permeate at pH 10 (lane 10). Therefore, highly purified plasmid DNA can be obtained from *E. coli* using a two-stage microfiltration process with adsorption and desorption.

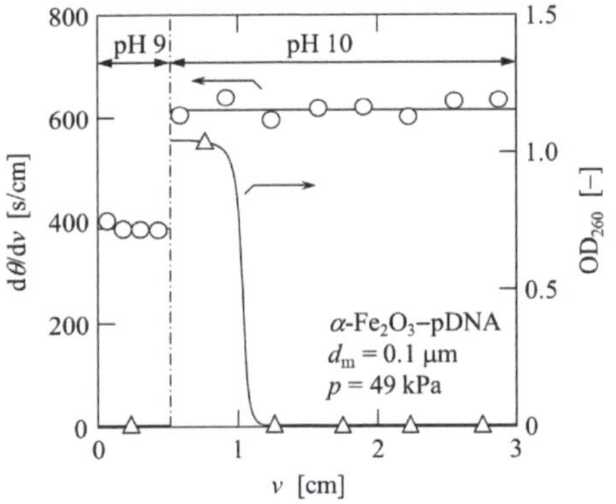

Figure 7. Variation of reciprocal filtration rate and optical density with filtrate volume per unit membrane area in liquid permeation (1 M Tris-HCl buffer pH 9 and 2 M Tris-HCl buffer pH 10).

3.3. Cell Disruption Properties

The above study applied the two-stage microfiltration process with adsorption and desorption and was performed on a solution containing plasmid DNA after alkaline lysis of *E. coli*. To establish a safe method that uses the minimum amount of chemicals possible, we investigated a plasmid DNA release method. The release of plasmid DNA from *E. coli* was attempted with ultrasonic irradiation, bead milling, and electroporation. In both methods, plasmid DNA was released from the *E. coli* suspension after treatment. However, long-term ultrasonic irradiation destroys the released plasmid DNA and does not increase the recovery amount, and bead mill disruption cuts the genomic DNA, rendering the subsequent purification difficult. Figure 8 shows micrographs of *E. coli* suspensions treated with each method. Compared with untreated cells, the change after electroporation was remarkable, and large flocs were formed. Biopolymers, such as genomic DNA, polysaccharides, and proteins, were released from *E. coli* using electroporation, and aggregates were formed along with the cells.

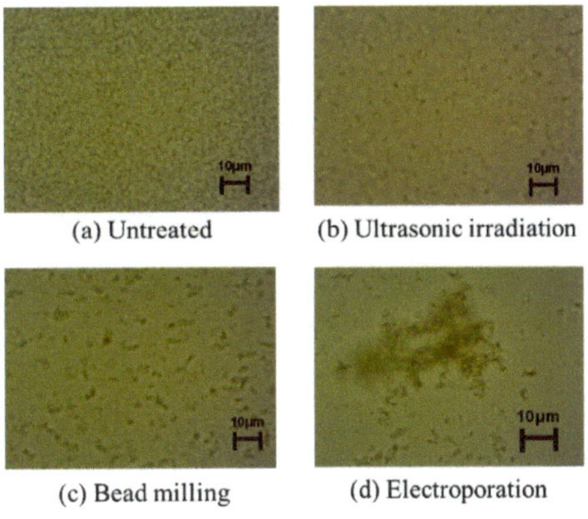

Figure 8. Photomicrographs of *E. coli* suspension treated under various conditions: (**a**) Untreated; (**b**) Ultrasonic irradiation; (**c**) Bead milling; (**d**) Electroporation.

Microfiltration was performed to remove impurities and obtain a plasmid DNA solution, and the results are plotted in Figure 9 in the form of the reciprocal filtration rate ($d\theta/dv$) against the filtrate volume v per unit effective membrane area. In the case of ultrasonic irradiation and bead milling, the flux decline was significant. In contrast, in the case of electroporation, in which aggregates were formed, the filtration rate was extremely high, confirming the superior separation performance. OD_{260} of the filtrates obtained using ultrasonic irradiation, bead milling, and electroporation were 8.80, 9.85, and 7.08, respectively, and OD_{260}/OD_{280} ratios were 1.5, 1.9, and 2.0, respectively. In terms of plasmid DNA extraction, Haberl et al. show that electroextraction leads to a higher concentration of extracted plasmid DNA than alkaline lysis, which is commonly used [22]. This filtrate containing plasmid DNA can be purified using a two-stage microfiltration process using α-Fe_2O_3, as described above. The recovered solution using a two-stage affinity microfiltration exhibited an OD_{260}/OD_{280} ratio of 1.8, indicating a high degree of nucleic acid purification [29]. However, to further improve the degree of purification of plasmid DNA, an operation to remove RNA, such as the addition of $CaCl_2$ or degradation with RNase, is required. In addition, attention should be paid to the detection and removal of impurities that do not contribute to the OD_{260}/OD_{280} ratio.

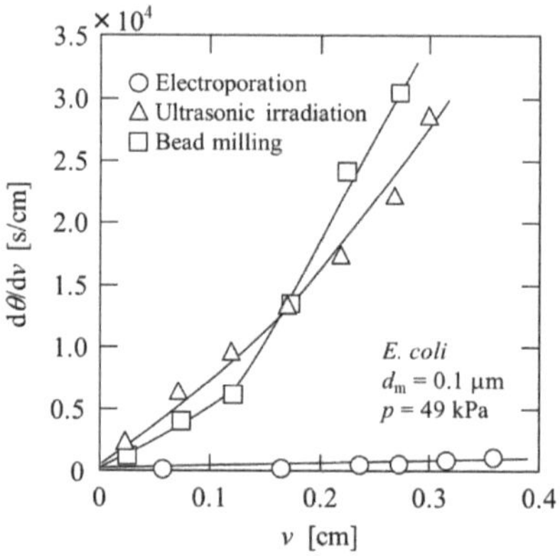

Figure 9. Microfiltration behavior of *E. coli* suspension treated under various conditions.

4. Conclusions

The affinity microfiltration of plasmid DNA using α-Fe_2O_3 as a ligand was examined. The adsorption and desorption properties of plasmid DNA revealed that α-Fe_2O_3 particles are suitable ligands. The data from two-stage affinity microfiltration, including both the binding process of plasmid DNA to α-Fe_2O_3 and the elution process of bound plasmid DNA, demonstrate that this method has immense potential for plasmid DNA purification. However, adsorption and desorption times should be optimized to reduce processing time. In addition, electroporation is effective as an elution method for bacterial cells in the purification process of plasmid DNA. The microfiltration performance was high owing to the formation of aggregates of impurities, including cells. Furthermore, the degree of nucleic acid purification was high. We believe that the results of this study will contribute to the establishment of a purification process suitable for the mass production of pharmaceutical-grade plasmid DNA.

Author Contributions: Conceptualization, N.K. and E.I.; methodology, N.K.; validation, D.S., T.S. and M.K.; formal analysis, D.S., T.S. and M.K.; investigation, D.S., T.S. and M.K.; writing—original draft preparation, N.K.; writing—review and editing, E.I. All authors have read and agreed to the published version of the manuscript.

Funding: This research was funded by JSPS KAKENHI, grant numbers JP 25420800 and 20K05190.

Institutional Review Board Statement: Not applicable.

Data Availability Statement: Not applicable.

Conflicts of Interest: The authors declare no conflict of interest.

References

1. Sayed, N.; Allawadhi, P.; Khurana, A.; Singh, V.; Navik, U.; Pasumarthi, S.K.; Khuana, I.; Banothu, A.K.; Weiskirchen, R.; Bharani, K.K. Gene therapy: Comprehensive overview and therapeutic applications. *Life Sci.* **2022**, *294*, 120375. [CrossRef]
2. Bohle, K.; Ross, A. Plasmid DNA production for pharmaceutical use: Role of specific growth rate and impact on process design. *Biotecnol. Bioeng.* **2011**, *108*, 2099–2106. [CrossRef]
3. Saade, F.; Petrovsky, N. Technologies for enhanced efficacy of DNA vaccines. *Expert Rev. Vaccines* **2012**, *11*, 189–209. [CrossRef] [PubMed]
4. Prazeres, D.M.F.; Monteiro, G.A.; Ferreira, G.N.M.; Diogo, M.M.; Ribeiro, S.C.; Cabral, J.M.S. Purification of plasmids for gene therapy and DNA vaccination. *Biotecnol. Annu. Rev.* **2001**, *7*, 1–30.
5. Prather, K.J.; Sagar, S.; Murphy, J.; Chartrain, M. Industrial scale production of plasmid DNA for vaccine and gene therapy. *Enzyme Microbiol.* **2003**, *33*, 865–883. [CrossRef]
6. Ghanem, A.; Healey, R.; Adly, F.G. Current trend in separation of plasmid DNA vaccines: A review. *Anal. Chim. Acta* **2013**, *760*, 1–15. [CrossRef]
7. Abdulrahman, A.; Ghanem, A. Recent advances in chromatographic purification of plasmid DNA for gene therapy and DNA vaccines: A review. *Anal. Chim. Acta* **2018**, *1025*, 41–57. [CrossRef]
8. Hirasaki, T.; Sato, T.; Tsuboi, T.; Nakano, H.; Noda, T.; Kono, A.; Yamaguchi, K.; Imada, K.; Yamamoto, N.; Murakami, H.; et al. Permeation mechanism of DNA molecules in solution through cuprammonium regenerated cellulose hollow fiber (BMM^{tm}). *J. Membr. Sci.* **1995**, *106*, 123–129. [CrossRef]
9. Morão, A.M.; Nunes, J.C.; Sousa, F.; Pessoa de Amorim, M.T.; Escobar, I.C.; Queiroz, J.A. Ultrafiltration of supercoiled plasmid DNA: Modeling and application. *J. Membr. Sci.* **1996**, *116*, 191–197. [CrossRef]
10. Kahn, D.W.; Butler, M.D.; Cohen, D.L.; Gordon, M.; Kahn, J.W.; Winkler, M.E. Purification of plasmid DNA by tangential flow filtration. *Biotechnol. Bioeng.* **2000**, *69*, 101–106. [CrossRef]
11. Kendall, D.; Lye, G.J.; Levy, M.S. Purification of plasmid DNA by an integrated operation comprising tangential flow filtration and nitrocellulose adsorption. *Biotechnol. Bioeng.* **2002**, *79*, 816–822. [CrossRef] [PubMed]
12. Latulippe, D.R.; Ager, K.; Zydney, A.L. Flux-dependent transmission of supercoiled plasmid DNA through ultrafiltration membranes. *J. Membr. Sci.* **2007**, *294*, 169–177. [CrossRef]
13. Higuchi, A.; Kato, K.; Hara, M.; Sato, T.; Ishikawa, G.; Nakano, H.; Satoh, S.; Manabe, S. Rejection of single stranded and double stranded DNA by porous hollow fiber membranes. *J. Membr. Sci.* **2011**, *378*, 280–289. [CrossRef]
14. Nunes, J.C.; Morão, A.M.; Nunes, C.; Pessoa de Amorim, M.T.; Escobar, I.C.; Queiroz, J.A. Plasmid DNA recovery from fermentation broths by a combined process of micro- and ultrafiltration: Modeling and application. *J. Membr. Sci.* **2012**, *415–416*, 24–35. [CrossRef]
15. Padilla-Zamudio, A.; Guerrero-Germán, P.; Tejeda-Mansir, A. Plasmid DNA primary recovery from *E. coli* lysates by depth bed microfiltration. *Bioprocess Biosyst. Eng.* **2015**, *38*, 1091–1096. [CrossRef] [PubMed]
16. Iritani, E.; Katagiri, N.; Kawabata, T.; Takaishi, Y. Chiral separation of tryptophan by single-pass affinity inclined ultrafiltration using hollow fiber membrane module. *Sep. Purif. Technol.* **2009**, *64*, 337–344. [CrossRef]
17. Hadik, P.; Szabo, L.P.; Nagy, E.; Farkas, Z. Enantioseparation of D,L-lactic acid by membrane techniques. *J. Membr. Sci.* **2005**, *251*, 223–232. [CrossRef]
18. Diogo, M.M.; Queiroz, J.A.; Prazeres, D.M.F. Chromatography of plasmid DNA. *J. Chromatogr. A* **2005**, *1069*, 3–22. [CrossRef]
19. Tan, L.; Kim, D.S.; Yoo, I.K.; Choe, W.S. Harnessing metal ion affinity for the purification of plasmid DNA. *Chem. Eng. Sci.* **2007**, *62*, 5809–5820. [CrossRef]
20. Han, Y.; Forde, G.M. Single step purification of plasmid DNA using peptide ligand affinity chromatography. *J. Chromatogr. B* **2008**, *874*, 21–26. [CrossRef]
21. Da Silva, N.R.; Jorge, P.; Martins, J.A.; Teixeira, J.A.; Marcos, J.C. Initial Screening of poly(ethylene glycol) amino ligands for affinity purification o plasmid DNA in aqueous two-phase systems. *Life* **2021**, *11*, 1138. [CrossRef] [PubMed]
22. Haberl, S.; Jarc, M.; Štrancar, A.; Peterka, M.; Hodžić, D.; Miklavčič, D. Comparison of alkaline lysis with electroextraction and optimization of electric pulses to extract plasmid DNA from *Escherichia coli*. *J. Membr. Biol.* **2013**, *246*, 861–867. [CrossRef]
23. Padilla-Zamudio, A.; Lucero-Acuña, J.A.; Guerrero-Germán, P.; Ortega-López, J.; Tejeda-Mansir, A. Efficient disruption of *Escherichia coli* for plasmid DNA recovery in a bead mill. *Appl. Sci.* **2018**, *8*, 30. [CrossRef]
24. Katagiri, N.; Kuwajima, Y.; Kawahara, H.; Yamashita, R.; Iritani, E. Special features of microbial cake under high pressure conditions in microfiltration. *Sep. Purif. Technol.* **2022**, *303*, 122234. [CrossRef]
25. Eon-Duval, A.; MacDuff, R.H.; Fisher, C.A.; Harris, M.J.; Brook, C. Removal of RNA impurities by tangential flow filtration in an RNase-free plasmid DNA purification process. *Anal. Biochem.* **2003**, *316*, 66–73. [CrossRef]
26. Katagiri, N.; Tomimatsu, K.; Date, K.; Iritani, E. Yeast cell cake characterization in alcohol solution for efficient microfiltration. *Membranes* **2021**, *11*, 89. [CrossRef]

27. Liu, C.H.; Tsao, M.H.; Sahoo, S.L.; Wu, W.C. Magnetic nanoparticles with fluorescence and affinity for DNA sensing and nucleus staining. *RSC Adv.* **2017**, *7*, 5937–5947. [CrossRef]
28. Ruth, B.F. Studies in filtration. III. Derivation of general filtration equations. *Ind. Eng. Chem.* **1935**, *27*, 708–723. [CrossRef]
29. Rogers, N.L.; Cole, S.A.; Lan, H.C.; Crossa, A.; Demerath, E.W. New saliva DNA collection method compared to buccal cell collection techniques for epidemiological studies. *Am. J. Hum. Biol.* **2007**, *19*, 319–326. [CrossRef]

membranes

Article

Screening and Scale-Up of Nanofiltration Membranes for Concentration of Lactose and Real Whey Permeate

Katrin Hofmann [1,*] and Christof Hamel [1,2]

1 Applied Biosciences and Process Engineering, Anhalt University of Applied Sciences, Bernburger Straße 55, 06366 Koethen, Germany
2 Institute of Process Engineering, Otto von Guericke University, Universitaetsplatz 2, 39109 Magdeburg, Germany
* Correspondence: katrin.hofmann@hs-anhalt.de; Tel.: +49-03496-672695

Abstract: In dairy industry huge quantities of whey accumulate as a by-product. In particular the containing lactose was not produced profitably in the past. Thus, the trend goes towards modification and sustainable use of lactose for which a concentration step is required. Nanofiltration (NF) has shown to be a good choice since partial demineralization can be realized in parallel. Therefore, in this study, 10 commercial polymer NF membranes were studied in detail and systematically for their suitability to concentrate lactose, with the proviso of high flux and high to complete rejection. Preliminary trials were conducted with flat-sheet membranes and a lactose model solution and the influence of transmembrane pressure (TMP), temperature and lactose concentration was studied. Finally, results were evaluated by using spiral wound modules and real industrial whey permeate. The results offered that a membrane screening is essentially since no correlation between molecular weight cut-off (MWCO) and permeate flow could be found. The conclusions found for the lactose model solution were in good agreement with the whey permeate, but as the ions contribute to the osmotic pressure of the feed the deviations increase in the course of concentration since ions are also partly retained.

Keywords: polymeric membrane; nanofiltration (NF); concentration of lactose and whey permeate; process parameters; flat-sheet module; scale-up; spiral wound module; molecular weight cut-off (MWCO)

Citation: Hofmann, K.; Hamel, C. Screening and Scale-Up of Nanofiltration Membranes for Concentration of Lactose and Real Whey Permeate. *Membranes* **2023**, *13*, 173. https://doi.org/10.3390/membranes13020173

Academic Editor: Daqi Cao

Received: 12 December 2022
Revised: 20 January 2023
Accepted: 26 January 2023
Published: 31 January 2023

1. Introduction

The utilization of whey has undergone major changes in recent years and decades. In order to produce 1 kg of cheese, around 10 L of milk are used and around 9 L of whey result as a by-product [1]. In the past whey was disposed of as waste in sewage and rivers, which led not only to environmental problems because of a high chemical oxygen demand (COD) between 50–60 kg/m^3 and high biological oxygen demand of around 30–40 kg/m^3 but also to a loss of valuable resources [1–3]. Besides legal regulations regarding waste water treatment and grown scientific knowledge of the value of whey components, the interest and awareness has raised with respect to increasing sustainability that are characterized by value chains with complete utilization of all milk components along with recycling product wastewater treatment [4,5]. In particular, the lactose as a main component of the side-product whey has not been used profitably in the past since the production has not been economically beneficial and the traditional lactose market is rather stagnating [4]. Additionally, around the world there is a great diversity in the distribution of lactose intolerance among adults ranking between 1% in the Netherlands and 98% in Southeast Asia [6]. Nevertheless, globally the dairy market is still growing because demand for dairy products in developing countries is growing along with income [7]. Therefore, the dairy industry is looking for alternative ways of lactose utilization. There are many approaches reported in literature, such as the synthesis of the prebiotics galacto-oligosaccharides, lactulose, lactitol, lactobionic acid, biogas or bioethanol [4,8,9] that require concentrated lactose to enhance reaction rates and synthesis efficiency, respectively [8,10]. Since the structure of dairy industry around the world is diverse, a mixture of many small family dairies and a few large industrial dairies [7], it is in general reasonable to concentrate the lactose prior to further processing or for cost-effective transport. In recent years nanofiltration (NF) as a powerful tool for concentration of lactose has gained raising interest because in parallel a partial demineralization can be realized that is often essential for subsequent processes [11,12]. Conventional demineralization strategies of whey include ion exchange and electrodialysis which are accompanied by high investment and operating costs [13]. In contrast, NF is, like reverse osmosis (RO) and ultrafiltration (UF), a pressure-driven membrane process and lies in between these two filtration processes in terms of operating pressure and separation limits and thus has corresponding characteristics of both RO and UF [14].

Previous studies have shown that NF is generally suitable for concentrating lactose (molecular weight = 342 g/mol) in whey permeate, with the positive side effect of partial demineralization compared to conventional evaporation, where all compounds within the feed are concentrated [3,11,12,15]. The focus of these studies was primarily on checking the feasibility of the process and investigating the influence of selected operating parameters whereas classification and comparability of the results is difficult because different membranes and/or operating parameters were used that influence the extent of demineralization. The impact of the choice of membrane was demonstrated by Räsänen et al. who compared four commercial NF membranes that varied in their solute retentions [12]. With rising volume concentration ratio (VCR) in the course of concentration the retention of ions slightly decreases and the degree of demineralization is increased [11,15]. The demineralization can be further enhanced through diafiltration (DF), which is accompanied by a greater loss of lactose [11]. Thus, DF is associated with a trade-off between the degree of demineralization and loss of lactose, whereas a mathematical model can support the optimization of the DF process as described in [16].

Therefore, the focus of this study is the comparison of ten commercially available membranes within the same experimental setup and reproducible experimental conditions. The aim is to identify the membrane with the best overall performance in terms of flux and retention in the sense of an ecological and efficient process and to get a better understanding of the extent of influence of different operating parameters.

Generally, it would be favorable to be able to model the membrane process in order to reduce the number of trials and thus the costs for the process design [17]. The experiments performed and the data derived provide a basis for that. However, the filtration process is based on complex interactions that determine the outcome of the overall process. The main influencing factors include the membrane properties, the composition of the feed and the operating parameters, which also interact with each other and also may affect significantly the three thermodynamic driving forces: concentration, pressure and voltage differences [18,19]. Nowadays the precise membrane composition is mostly a trade secret, since the membrane manufacturers are very secretive about the exact composition and structure of the membranes [20,21]. This in turn makes predictions even more difficult, especially in the context of interactions between membrane and complex feeds. Also, because structural properties of the pores and the solute, like size, shape and length and physicochemical properties like zeta potential and hydrophilicity influence the permeation behavior [22]. The separation mechanism in NF is based on charge and size exclusion [22,23]. Flow and selectivity of neutral solutions in NF can be described reasonably well through the solution-diffusion model originally formulated for RO. In contrast, the complexity of ionic solutions is very pronounced, since interactions with the membrane as well as with each other take place depending among other things on pH, temperature and concentration [14,24,25]. Various research approaches were pursued with single salts as well as with salt mixtures under different process conditions like pH and temperature in order to gain a better understanding of the interactions and extent of influence [14,23,26–28]. However, due to the complexity of the interactions involved in the nanofiltration of electrolyte solutions, the membrane performance cannot be fully predicted and further research is needed, especially as complex mixtures are a great challenge [14,17,28]. Actually, real industrial whey is a rather complex mixture of compounds, many of which may interfere with the membrane surface, or compounds that could interact with each other in different ways. A further challenge is that there is not just one type of whey but that the composition is diverse depending on seasonal variation within the raw milk but also due to different manufacturing, production and processing processes [5,13]. Thus, preliminary trials are still essentially for an overall economically and ecologically process design, ideally with the desired process parameters and feed [14,17]. Therefore, in terms of systematics, first orienting trials were carried out with a model solution consisting of pure lactose to initially exclude the complex interactions caused by the whey permeate matrix with its ions, acids and other components like amino acids, peptides and proteins. Finally, these data were compared with the performance when industrial whey permeate had been used.

In summary, this study has three main objectives:

1. Membrane screening—Out of 10 membranes, identification of those membranes with the best performance in terms of high flux and high retention of lactose.
2. Determination of process parameter settings and its extent of influence for achieving high filtration fluxes and high retention of lactose.
3. Scale-up and evaluation of the influence of industrial whey permeate on the course of filtration compared to the lactose model system.

2. Materials and Methods

2.1. Feedstock

All feed solutions were preserved from microbiological spoilage by addition of 0.02% (*w*/*w*) sodium azide (Carl Roth, Karlsruhe, Germany) in order to avoid the formation of biofilms that could affect the results. For preliminary experiments a model solution was used that was prepared by dissolving a pharmaceutical-grade α-lactose monohydrate (Carl Roth, Karlsruhe, Germany) in deionized water in the amounts needed to reach the desired lactose concentrations of approximately 25, 50 and 75 g/L.

Fresh sweet whey permeate was taken from a feed line of an industrial filtration process of a local dairy plant where it had been separated by a 10 kDa UF membrane and was stored at 4 °C until use. The average composition of the sweet whey permeate used is shown in Table 1.

Table 1. Average composition of the sweet whey permeate used.

	Sweet Whey Permeate
pH	6.5
conductivity (mS/cm)	3.7
Dry matter (%)	3.2
Lactose (g/L)	26.3 (→ 25 for lactose model solution)
Protein (%)	0.1
Sodium (mg/L)	169.5
Potassium (mg/L)	776.8
Ammonium (mg/L)	70.1
Magnesium (mg/L)	29.9
Calcium (mg/L)	193.0
Lactic acid (mg/L)	376.6
Chloride (mg/L)	464.5
Nitrate (mg/L)	45.6
Phosphorus (mg/L)	606.0
Sulfate (mg/L)	60.6

2.2. General Experimental Setup of Nanofiltration Trials

The general structure and/or flow sheet of the semi-pilot scale filtration plant used for the experiments was provided by inocermic (Hermsdorf, Germany) is shown in Figure 1.

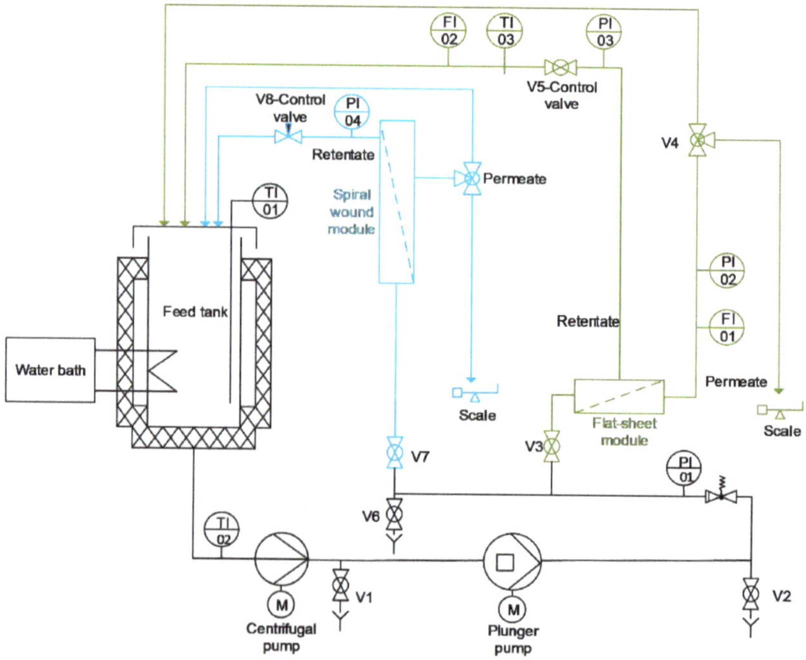

Figure 1. Piping and instrumentation diagram of filtration plant (green—flat-sheet module circuit; blue—1812″ spiral wound module circuit).

The plant was designed for tests in crossflow mode and was equipped with both a pressure housing for flat-sheet membranes and one for 1812″ spiral wound modules realizing scale-up. Figure 2 illustrates the construction of the flat-sheet test cell. The feed flew parallel to the membrane and the retentate was recirculated perpendicularly to the membrane surface.

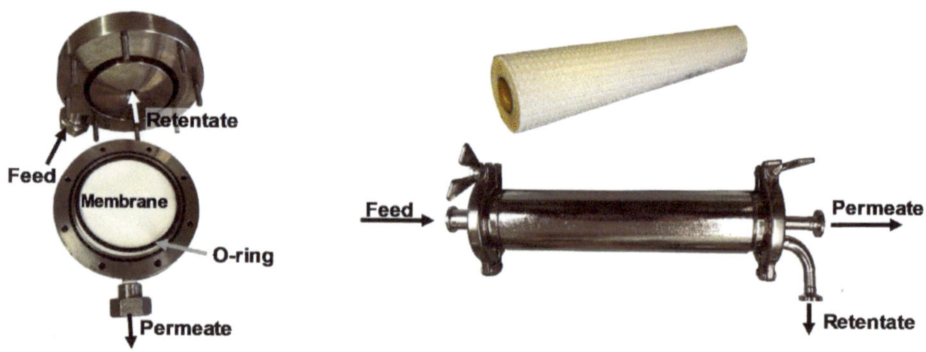

(a) (b)

Figure 2. Visualization of modules used. (**a**) flat-sheet module; (**b**) spiral wound membrane and housing (1812″).

All pipes and the feed tank were insulated and temperature was kept constant during operation using a water bath (F32, Julabo, Seelbach, Germany) that was connected to the double jacket of the feed tank. The actual temperature in the feed tank was measured directly by the digital temperature sensor TI-01 in the feed using a thermocouple. The maximum capacity of the feed tank was 12 L.

The plant had a centrifugal pump (CDX/A 70/05, EBARA Pumps Europe, Gambellara, Italy) that acted as a feed pump for the high pressure plunger pump (271 D1110, Cat Pumps, Idstein, Germany) which was responsible for the flow and pressure generation during the process. Both pumps were controlled via frequency converters in order to set the desired volume flow. For the flat-sheet trials a volume flow of 0.30 m^3/h had been used while with the spiral wound module a volume flow of 0.43 m^3/h had been set. The retentate-side pressures were adjusted manually to the desired value by the control valves. In case of the flat-sheet module the control valve V5 was used and the pressure was measured by the pressure gauge PI-03. For the spiral wound module the transmembrane pressure (TMP) was adjusted by the needle valve V8 and monitored by the pressure gauge PI-04.

The permeate weight for calculating the permeate flux was monitored by a scale (Scout STX6201, OHAUS Europe, Greifensee, Switzerland).

2.3. Filtration Test Scheme

2.3.1. Membranes and Experimental Matrix

In this study 10 commercially available NF membranes from 3 different manufacturers were investigated. Table 2 summarizes the membranes used within this study and gives an overview of their properties in terms of specified molecular weight cut-off (MWCO), maximum operating pressure and temperature as well as membrane material provided by the manufacturers.

Table 2. Overview of the nanofiltration membranes used in this study.

Model	Manufacturer	MWCO [1] (g/mol)	Ø MWCO [1] (g/mol)	Membrane Chemistry [2]	Max. Operating Temperature (°C)	Max. Operating Pressure (MPa)	
TS 50	MANN+ HUMMEL	200–300	250	Thin-Film Polypiper-azine	45	4.1	
TS 40		200–300	250				
XN45		300–500	400				
NP030		500–600	550	PES TFC	50	4.0	
NFS	Synder Filtration	100–250	175	Proprietary PA TFC	50	if T < 35 °C: 4.1	if T > 35 °C: 3.0
NFX		150–300	225				
NFW		300–500	400				
NDX		500–700	600				
NFG		600–800	700				
SR3D	KOCH Membrane Systems	200	200	TFC PA	45	4.5	

[1] MWCO = molecular weight cut-off; [2] PES = polyethersulfone; TFC = thin-film composite; PA = polyamide.

The experimental matrix had 3 stages and is summarized in Table 3. All trials were carried out at least in duplicate. At the beginning of every trial the actual membrane and its performance was checked and characterized via pure water flux measurements at different TMP steps from 0.5 MPa up to maximum 3.5 MPa in 0.5 MPa steps at the temperature of the respective trial. All water used in the filtration experiments was deionized water with 0.02% (w/w) sodium azide. The water flux measurements were also used to evaluate the cleaning efficiency.

Table 3. Overview of parameter settings and membrane used for experimental setup.

Stage	Membrane Module	Filtration Mode	Tested Membranes	Feed	T (°C)	TMP [1] (MPa)	Initial Feed-$c_{(Lac)}$ (g/L)
TMP-Screening	Flat-sheet	recirculation	all	lactose model solution	20	0.5→3.5→0.5	25
Parameter studies	Flat-sheet	concentration	TS40; XN45; NFS; NFG; SR3D	lactose model solution	20	2	25
						3	
					45	2	
			XN45; SR3D; TS40			3	25
			XN45; SR3D				50
							75
			NFS; NFG; TS40			2	50
							75
Scale-up	1812″ spiral wound	concentration	NFS	lactose model solution	45	2	25–75
				whey permeate			
			XN45	lactose model solution	45	3	
				lactose model solution	20	2	
				whey permeate			

[1] TMP = transmembrane pressure.

The first stage started with a TMP-screening of all 10 membranes in order to be able to compare the performance of the membranes in terms of retention and permeate flux. Based on these results, in the second step (parameter studies) a pre-selection of 5 membranes was carried out, in which the influence of various process parameters was examined in more detail.

These first 2 test stages were carried out with flat-sheet membranes that were delivered as dry sheets and cut to a circular diameter of 9.1 cm to fit the test cell. After subtraction of the O-ring area, this results in an effective filtration area of 0.0053 m^2. At the last stage (scale-up), experiments were carried out with 1812″ spiral wound modules, concrete with the NFS from Synder Filtration (Vacaville, CA, USA) with a filtration area of 0.28 m^2 and the XN45 from MANN+HUMMEL (Ludwigsburg, Germany) with a filtration area of 0.20 m^2. Both spiral wound modules had a feed spacer thickness of 46 mil.

2.3.2. First Stage—TMP-Screening

Prior to every experiment the water flux was measured as explained in Section 2.3.1. The TMP-screening was conducted at 20 °C with 5 kg of lactose model solution with a lactose concentration around 25 g/L which was based on the initial lactose concentration of the industrial sweet whey permeate. In order to maintain a constant composition of the feed, the permeate was regularly returned to the feed tank which means the experiments were done in recirculation mode. The TMP was increased in steps of 0.5 MPa, starting with 0.5 MPa to maximum of 3.5 MPa or the maximum operating pressure of the membrane followed by a stepwise decrease of the TMP to prove hysteresis. At every investigated TMP after at least 10 min at this TMP or when the permeate flux was stable, the permeate flux was measured and a sample of the feed and the permeate was taken for lactose quantification via HPLC (Section 2.4.1) in order to calculate the lactose retention.

After each experiment the plant and the membrane was firstly thoroughly flushed pressureless with tap water at 45 °C followed by deionized water at room temperature. In order to profoundly clean the flat-sheet test cell, it was disassembled and the membrane was carefully rinsed with deionized water and stored in deionized water with 0.02% sodium azide inside the reassembled test cell. Water flux measurements confirmed the successful cleaning procedure.

2.3.3. Second Stage—Parameter Studies

With the 5 selected membranes (see Table 3), parameter studies were carried out in concentration modes with 7 kg of lactose model solution, i.e., after 10 min with recirculating permeate, the feed was concentrated to a VCR of 2.25. As a starting point for the comparative evaluation of the influence of various process parameters the following standard process parameters were set: ϑ = 20 °C, TMP = 2 MPa and an initial lactose concentration of 25 g/L. The permeate flux was measured at defined VCR (every 0.25 step) and every 30 min. At these defined VCR there were also samples taken for lactose quantification from the feed, the permeate out of the permeate pipe and from the cumulating permeate. The experiments were conducted at 45 °C to investigate the influence of the temperature. The influence of the TMP was investigated at 3 MPa. The combined effect of temperature and TMP was studied at 45 °C and 3 MPa with those membranes that tolerated the combination of these process parameters. Finally, the impact of increasing initial lactose concentration within the feed was studied in detail. Because of the limiting maximum tank volume, the experiments were divided in separate experiments with an initial lactose concentration of 50 g/L and 75 g/L. Depending on the tolerance limits of the membranes those trials were carried out either at 45 °C and 2 MPa (membranes from Synder Filtration) or at 45 °C and 3 MPa (the other membranes). The cleaning procedure was identical as for the TMP-screening.

2.3.4. Third Stage—Scale-Up

The scale-up was carried out with those 2 membranes that had revealed the best performance at the previous stages. The experiments were conducted at different process variables for each membrane, because of differences in the maximum operating pressure at temperatures above 35 °C. In order to compare the performance of flat-sheet membranes with spiral wound membranes and evaluate the potential of experiments based on flat-sheet membranes, firstly the lactose model solution was concentrated. In case of the NFS from Synder Filtration the concentration has been performed at 45 °C and a TMP of 2 MPa while with the XN45 from MANN+HUMMEL the concentration was carried out at 45 °C and 3 MPa as well as 20 °C and 3 MPa. In a second step the lactose in sweet whey permeate was concentrated with the same process parameters for the NFS and at 20 °C and 3 MPa for the XN45.

The general procedure of the concentration trials was as followed: As it was intended to reach approximately 200 g/L lactose within the concentrate and the initial lactose concentration was given by the amount in the sweet whey permeate, the maximum tank volume was not sufficient for one whole process. Therefore, the trials started with the maximum feed volume of 14.5 L that was recirculated for 10 min and then the concentration was carried out until a VCR of 2.5 was reached. This procedure was repeated 3 times and the resulting concentrate of all 3 runs was united and 14.5 L of this retentate were concentrated to a VCR of 3, resulting in a total VCR of 4.5. At defined VCR (every 0.5 step, except at the beginning additionally at 1.25) the permeate flux was measured and samples were taken for lactose quantification from the feed, the permeate out of the permeate pipe and from the cumulating permeate. When the whey permeate was used as feed, the conductivity and ion composition of the token samples was additionally analyzed.

The solute rejection (R_j) of the lactose and the ions points to the percentage of solute that does not pass the membrane and was calculated based on the following Equation (1) [29].

$$R_j(\%) = \left(1 - \frac{c_{j,p}}{c_{j,f}}\right) \times 100 \tag{1}$$

where $c_{j,p}$ and $c_{j,f}$ are the solute concentration in the permeate pipe and the feed concentration at a defined VCR. The loss respectively removal of solute was calculated based on balancing the masses of the solute in the feed and in the cumulating permeate as follows:

$$Loss_i(\%) = \frac{m_{i,\,p}}{m_{i,\,f}} \times 100\% \tag{2}$$

where $m_{i,\,p}$ is the mass of the solute in the cumulative permeate and $m_{i,f}$ is the mass of the solute in the initial feed.

When working with the lactose model solution the cleaning procedure was the same as for the TMP-screening. In case whey permeate was used as feed, the cleaning procedure was extended by an additional cleaning step with an enzymatic cleaning agent (Ultrasil 53, Ecolab, Monheim am Rhein, Germany) at a concentration of 1% after the system had been flushed thoroughly with tap water. The enzymatic cleaning was performed pressureless at approximately 40 °C under constant recirculation of retentate and permeate for 30 min. Afterwards the plant was repeatedly flushed with tap water and then the previously described cleaning routine followed.

2.4. *Analysis Methods*

2.4.1. Lactose Quantification

Quantification of lactose concentration before and during the filtration process was realized via HPLC Chromaster® HPLC system (VWR, Darmstadt, Germany). The column Vertex Plus Eurokat Na, 300 × 8 mm ID with a 30 × 8 mm ID precolumn, particle size 10 µm (Knauer GmbH, Berlin, Germany) was used with a flow rate of 0.25 mL/min, ultrapure water with 0.02% sodium azide (*w/v*) as eluent and a sample injection volume of 10 µL. The column was kept at a temperature of 85 °C and the detection was carried out at 40 °C with a refractive index detector.

2.4.2. Quantification of Ions and Conductivity

Ions were analyzed by the ion Chromatograph DX-100 (Dionex Corporation, Sunnyvale, CA, USA), equipped with a conductivity detector. All samples were filtered through a 0.2 µm syringe filter before analysis, if appropriate diluted with deionized water and 25 µL of the sample were injected and measured at room temperature. For anions the column Dionex™ IonPac™ AS14 (4 × 250 mm) with precolumn Dionex™ IonPac™ AG14 (4 × 50 mm) was used with a flow rate of 1 mL/min, 3.5 mM Na_2CO_3 and 1 mM $NaHCO_3$ as eluent. Cations were analyzed using a Dionex™ IonPac™ CS 12A (4 × 250 mm) column with a Dionex™ IonPac™ CG14 (4 × 50 mm) precolumn, by using a mobile phase of 20 mM methanesulfonic acid, with a flow rate of 1 mL/min.

The conductivity was measured with a WTW conductimeter LF 539 (WTW, Weilheim, Germany) using nonlinear temperature compensation mode based on the reference temperature of 25 °C.

3. Results and Discussion

3.1. *Screening and Preselection of Membranes*

Within the scope of this study, NF membranes were to be examined with regard to their suitability for concentrating lactose (molecular weight (MW) = 342 g/mol). When searching for appropriate membranes the MWCO was used as the most important criterion for the selection, but it was found that the specifications in this regard varied greatly. In contrast to UF or microfiltration (MF) membranes, in the majority of cases ranges were given, e.g., for the XN45 300-500 Da and only for the membrane from KOCH Membrane Systems there was one value specified as MWCO. The given number of the MWCO expresses in general the molecular weight of a compound that is rejected to 90% by the given membrane [29,30]. However, one challenge concerning the MWCO is, that there exists no binding definition or methodology and manufacturers may use different methods and feed solutions for determination [30]. Therefore membranes with a higher MWCO than the MW of lactose were also included in the investigations. Additionally this approach was based on the following assumptions: On the one hand, it was supposed that higher MWCO tended to be associated with higher fluxes [3,23], which could result in an overall increase in process efficiency and profitability, although this might entail some losses of lactose. On the other hand, whey permeate contains a large number of compounds and minerals besides lactose [3,5,13]. Therefore, when using membranes with a larger MWCO, the demineralization could possibly also be promoted or potential interfering components that could impair the flux might unfold their effect less strongly. For that reason, as a first step, it was necessary to examine the eligible membranes with regard to their retention capacity for the target compound lactose and the resulting flux. This was realized as an important part of a TMP-screening.

In order to check the reproducibility of the results, tests were carried out with the same membrane as well as with several sections of the same provided membrane lot and, in the case of the NFG, with cuts from different lots. This strategy ensures the testing of a representative sample. In the case of membranes that were used repeatedly and those from the same batch, the curves basically corresponded very well, so that good reproducibility of the tests could be assumed. The data shown represent the mean values of the respective tests.

As shown in Figure 3 using the XN45 as an example, all membranes, with the exception of the NFG, showed basically the same course of the curve during the TMP screening, i.e., a stepwise increase of the retentate-side pressure led to a linear increase in flux. This indicates that the permeate flux is a result of the membrane resistance since the process is not mass transfer limited [31]. These observations are in good agreement with the results of [11,15], whereas the studies of [3,15] demonstrate that mainly remaining proteins in whey are responsible for critical flux phenomena during NF, where a further increase in TMP does not result in an equivalent increase in mass flow through the membrane.

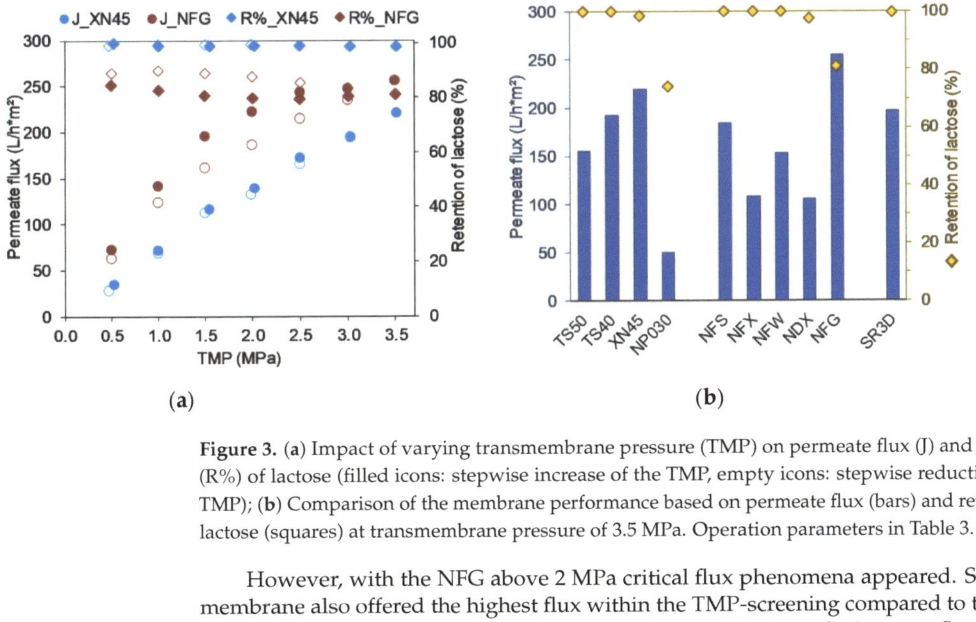

(**a**) (**b**)

Figure 3. (**a**) Impact of varying transmembrane pressure (TMP) on permeate flux (J) and retention (R%) of lactose (filled icons: stepwise increase of the TMP, empty icons: stepwise reduction of the TMP); (**b**) Comparison of the membrane performance based on permeate flux (bars) and retention of lactose (squares) at transmembrane pressure of 3.5 MPa. Operation parameters in Table 3.

However, with the NFG above 2 MPa critical flux phenomena appeared. Since this membrane also offered the highest flux within the TMP-screening compared to the other membranes, it was assumed that the support layer might impede the mass flow, so that a further increase in pressure does not lead to a further linear increase in flux. Chen et al. investigated the effects of different fabrication parameters in the course of the production of a home-made PES NF spiral wound module. Their work demonstrated that with increasing PES-substrate thickness the pure water permeability decreases [21]. An effect through components of the feed solution could be excluded since the results with the other membranes showed that with the investigated lactose concentration no concentration polarization occurred. Furthermore, in the course of the gradual increase and decrease of the TMP a hysteresis effect was found for the NFG and the NP030, i.e., when decreasing the TMP the same flux as before was not reached but the flux was further reduced, as can be seen in Figure 3. In parallel, this led to an increased retention of lactose. Since this behavior only occurred with two of the ten membranes examined, fouling or the formation of a concentration polarization layer due to the feed can be excluded, since otherwise all curves would have to be similar. For the NP030, the compaction as a result of high pressures is known by the manufacturer, i.e., the application of high pressure leads to a compaction of the membrane that is irreversible and thus modifies the membrane performance. This could also be confirmed in repeated experiments with the same membrane. Lower fluxes were always measured in the second run. In contrast, the curves of two consecutive trials with the same NFG membrane were in very good agreement. Therefore, it was assumed that generally compaction does occur, since it shows comparable curves to NP030, but that this is reversible, and might be due to more elastic membrane material components.

Due to the agreement of the curves of the flux as a function of the TMP, it was possible to compare all membranes at one specific TMP in order to identify those membranes with the highest fluxes. In Figure 3 all ten membranes were compared by plotting the permeate flux and the rejection at 3.5 MPa.

Based on the MWCO specified by the manufacturers, there was an expectation with regard to the retention, i.e., that the retention would deteriorate with increasing MWCO, particularly if the MWCO was greater than the MW of lactose with 342 g/mol. Additionally, it was assumed that a higher MWCO would tend to be associated with a higher flux, since the mass flow would be less impeded by increasingly larger pores [3,23]. As a result, the following theoretical order of the membranes was derived from the MWCO specified by the manufacturers: (1)-NFS (175 Da) < (2)-SR3D (200 Da) < (3)-NFX (225 Da) < (4)-TS40 & TS50 (250 Da) <(6)-NFW & XN45 (400 Da) < (8)-NP030 (550 Da) < (9)-NDX (600 Da) < (10)-NFG (700 Da). In contrast, a significantly different order was found when the maximum flux was used for arrangement: (10)-NFG (700 Da) > (6)-XN45 (400 Da) > (2)-SR3D (200 Da) > (4)-TS40 (250 Da) > (1)-NFS (175 Da) > (4)-TS50 (250 Da) > (6)-NFW (400 Da) > (3)-NFX (225 Da) > (9)-NDX (600 Da) > (8)-NP030 (550 Da).

These results elucidate that it is not possible to derive performance with regard to flux based on the given MWCO and that the MWCO is at best indicative. In addition, the membrane manufacturers provide only little information about the membrane material and the exact structure of the support layer [20], so that often only general information are available (Table 2) and therefore these information cannot be used in detail for membrane selection either. This underlines the need for specific, system dependent membrane screening in order to identify the most suitable membrane with the highest flux for the specific filtration task.

The measured retentions were generally in agreement with the expectations, i.e., a MWCO < MW of lactose was associated with a complete retention, while a MWCO > MW of lactose resulted in a partial permeation of lactose. However, the extent of the lactose loss was different in comparable MWCO ranges. For example, in average the NDX (500–700 Da) rejected 97.4% of the lactose compared to the NP030 (500–600 Da) with only 73.7% lactose retention. These results also point out the necessity of a membrane screening, since membrane manufacturers use different methods and substances to determine the MWCO [30]. Thus, it might occur that a membrane with a supposedly not fully suitable MWCO specified by the manufacturer can still have a very good retention and possibly even convince in terms of overall performance (high flux + high retention) for a specific examined separation process near the given MWCO. Finally, the process design depends on various factors, and in particular significantly on the total process time which is affected by the flux. Therefore, those five membranes with the highest fluxes were selected for further investigations, even if complete retention was not achieved with all of these candidates.

3.2. *Effect of Process Parameters*

In the second part of this study the objective was to examine the influence of different process parameters on permeate flux and lactose retention, namely temperature, TMP, the combination of temperature and TMP and increasing lactose concentrations in the feed. Within this context it was particularly important to obey the allowable operating limits, which meant that the combined effect of increased temperature and TMP could not be examined with the membranes from Synder Filtration as these were not as pressure-stable as necessary (3 MPa) at 45 °C.

The results are summarized in Figure 4. The influence of the studied parameters was compared with the curves resulting from the concentration of 25 g/L lactose at 20 °C and 2 MPa using those data sets as standard parameters. The NFS, SR3D and XN45 showed very similar flux values in the course of concentration. Only the flux of the TS40 was lower and the NFG showed in agreement with the TMP-screening the highest fluxes. For the NFG this was accompanied by a reduced lactose retention, which improved to about 92% in the course of concentration. The retention of the other four membranes remained at a consistently high level of ≥97%.

Overall, it can be stated that with the exception of the NFG, the magnitude to which the membrane performance responds to a change in a process parameter was very similar for all other four membranes considered.

In several studies the effects of temperature on permeate flux and retention of solutes were investigated. Temperature can influence the membrane morphology and the properties of the feed with its solutes which in turn determine the membrane performance [31,32]. In general, an increase in temperature reduces the viscosity of the feed and thus results in an increased permeate flux [3,15,22,23,32]. While for UF membranes these observations can just be explained by the Hagen-Poiseuille law and show a linear relationship [3], for NF membranes with much smaller pores the underlying mechanisms are much more complex, affecting water permeability and rejection of solutes [22,32]. Tsuru et al. [22] have studied these phenomena extensively. They have carried out the investigations with inorganic ceramic membranes in order to exclude temperature effects on the membrane morphology since it is assumed that ceramic materials do not change when exposed to a temperature increase compared to polymeric materials. As a result they have proposed three potential explanations induced by increased temperatures: (a) The mass transport of water molecules through the micropores may be provoke an activated process promoted through the increase of thermal energy of the water molecules. (b) An increase of the effective pore diameter, because besides the given pore diameter the effective pore diameter is determined by the amount and layer thickness of adsorbed water molecules on the pore walls and this layer decreases with temperature rise. (c) The influence of temperature on the viscosity differs between the inside of the micropores and the bulk solution so that the viscous effect is more pronounced inside the pores [22]. Within this study a similar factor was found for all examined membranes. Increasing the temperature from 20 °C to 45 °C led to a doubling of the permeate flux for the NFS, NFG and TS40. In contrast, the flux of the SR3D increased only 1.7-fold and for the XN45 1.8-fold. The differences between the membrane performances with regard to the retention of the lactose were much more distinct. Precisely, for the NFG with the highest MWCO, an increase in temperature was associated with a decrease in rejection of 15–20%. However, the XN45, whose specified MWCO lay also above the MW of lactose, showed only a slight reduction in lactose retention of about 1.8%. For the other three membranes, an increase in temperature had no effect on the rejection. These differences between the membranes can only be explained by temperature effects on the polymer structure of the membrane material. Generally, both pore-widening and pore-narrowing effects may occur. A temperature increase might enhance the mobility of the membrane polymer chains which in turn might result in an additional polymer relaxation [27]. This might affect the effective membrane thickness and effective pore diameter, which in turn might influence the retention of the components [32]. Yao et al. [31] have determined the MWCO of four polymeric membranes at 20 and 50 °C with a solution of polyethylene glycols of different molar masses. At elevated temperatures they measured higher MWCO and concluded that the temperature rise induces changes in the polymeric membrane structure to the same extent for the open and tight membranes. However, temperature cycling experiments (T = 20 °C→T = 50 °C→T = 20 °C) revealed that these changes were only completely reversible for the tight membranes whereas the membrane performance of the more open membranes was altered due to permanent reorientation of the polymer structure [31]. The water flux measurements before and after the trials of the tested membranes within this study showed comparable curves indicating that no permanent changes in the membrane morphology due to temperature occurred. Several studies reported a rejection drop at elevated temperatures [3,33,34]. Precisely, for uncharged solutes like lactose and fructose an increased temperature was accompanied by a decrease in retention of those solutes. This was explained by the reduced viscosity and thus increased diffusivity due to higher temperatures. As a result, the permeation of neutral solutes was amplified [3,33,34]. Furthermore, changes in the membrane structure have to be considered. Ben Amar et al. [32] investigated the retention of four neutral solutes with MW between 92 and 342 g/mol by a polymeric NF membrane at elevated temperatures. Generally, for all four solutes the rejection decreased as the temperature arose. However, the strongest rejection drop occurred for the intermediate solutes whereas the temperature increase had only a minor impact on the sucrose with a MW of 342 g/mol [32]. These observations might among others explain the differences between the investigated membranes. The decrease in retention was most pronounced for the NFG with the highest MWCO, followed by the XN45 whose MWCO ranges partly above the MW of lactose. It can be deduced from this that it is important to conduct trials at elevated temperatures since the manufacturer's data on MWCO are usually gained at room temperature and thus the values and as a result the membrane performance at higher temperature might differ [31].

(**a**) standard parameter

(**b**) temperature

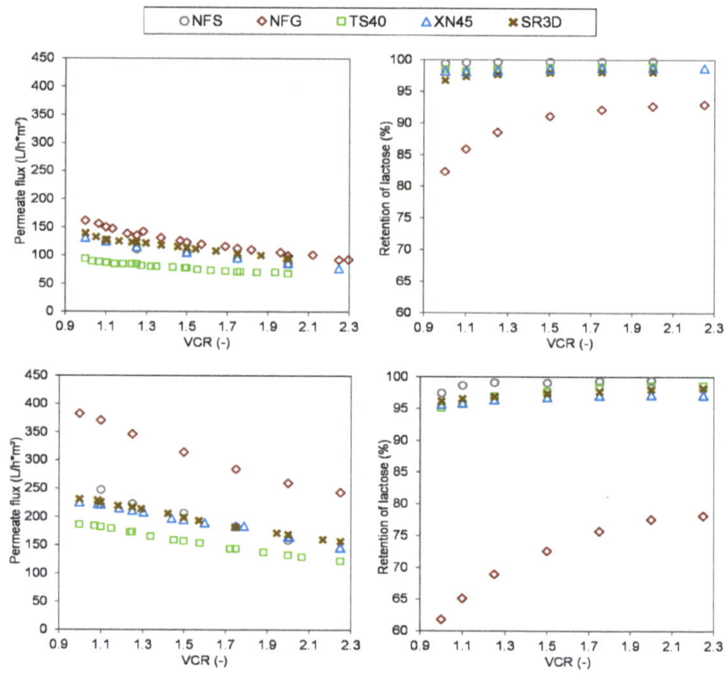

Figure 4. *Cont.*

(**c**) TMP

(**d**) temperature & TMP

(**e**) initial concentration of lactose at 45 °C & 2 MPa

(**f**) initial concentration of lactose at 45 °C & 3 MPa

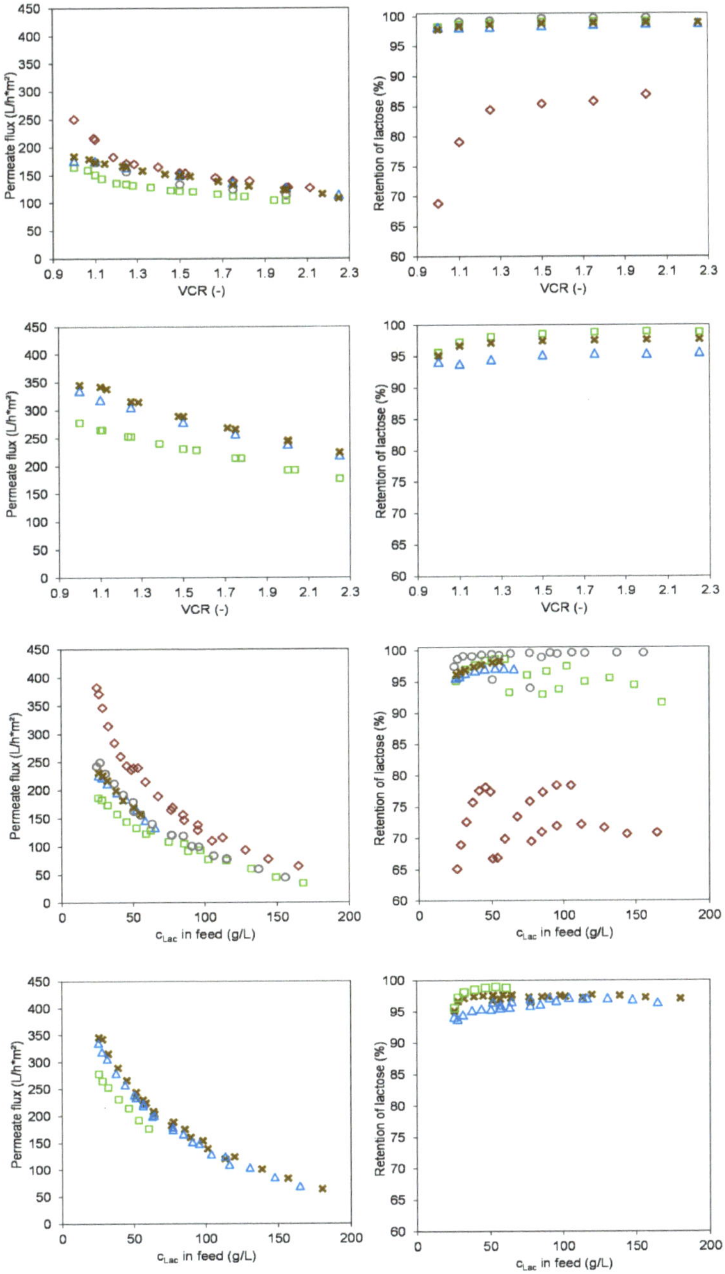

Figure 4. Course of permeate flux (left side) and retention of lactose (right side) during concentration of lactose in dependency of varying process parameters; (**a**) standard parameter (20 °C; 2 MPa; 25 g/L lactose); (**b**) temperature (45 °C); (**c**) transmembrane pressure (TMP) (3 MPa); (**d**) temperature and TMP (45 °C & 3 MPa); (**e**) initial concentration of lactose in feed (45 °C; 2 MPa; 25–75 g/L initial feed-c); (**f**) initial concentration of lactose in feed (45 °C; 3 MPa; 25–75 g/L initial feed-c). Operation parameters summarized in Table 3.

As already observed in the TMP-screening, an increase in TMP leads to a higher permeate flow rate, although the magnitude is less pronounced than the influence of temperature. In general, a pressure increase will at last lead to a limiting pressure where a further increase will not enhance the permeate flux any further. This is among others explained by the osmotic pressure model since the application of higher pressure leads to a further increase of the concentration of the solute near the membrane that is higher compared to the bulk concentration. As a result the increased concentration at the membrane surface reduces the driving force (pressure difference) and thus limits the permeate flux [35]. The TS40 responds most strongly to an increase in TMP (1.6-fold), the effect is least pronounced with the NFG. This supports the hypothesis formulated during the TMP-screening, that the membrane structure of the NFG might impede mass flow. For the NFG the increased TMP also has an adverse effect on retention. The TMP increase from 2 to 3 MPa results in an initial reduction in retention of approximately 13%, with a final difference of 7% at the VCR of 2.25. For the other four membranes, retention remains at a consistent high level.

As the three membranes SR3D, TS40 and XN45 withstand the combination of increased temperature and increased TMP, the combined effect of both parameters could be investigated. It was shown that the combination results in the greatest influence on the permeate flux as expected. With the TS40, the flux triples without changes in retention. For the XN45, the temperature and TMP increase results in a 2.7-fold increase in flux and only a slight decrease in retention of approximately 3.6%. The rejection of the SR3D is unchanged and the flux increased by a factor of 2.6.

In the last part of the parameter study, lactose was concentrated in three stages from 25 g/L to approximately 170 g/L. These experiments were carried out based on the results of the previous trials, i.e., since the temperature within the examined range had the greatest flux-improving effect without any significant reduction in retention, the process was realized at 45 °C (Table 3). Additionally, the solubility of lactose at 20 °C is rather low (15.9 wt%) compared to other sugars which could impede the concentration process. With rising temperatures there is a strong increase in solubility so that from this point of view it is favorable to work at as high temperatures as possible [36]. Since the membranes from Synder Filtration (NFS; NFG) do not have such a high pressure stability at high temperatures, the lactose was concentrated at 2 MPa while with the other three membranes at 3 MPa. The results are summarized in (e) and (f) of Figure 4. Generally, the curves of all membranes are similar. As expected, in the course of concentration the flux continuously decreases due to an increase in osmotic pressure, as described by several authors [11,12,15]. Di Giacomo et al. [36] have shown that the osmotic pressure is not much influenced by the temperature but strongly depended upon the concentration. Thus, the osmotic pressure determines the maximum possible concentration that can be reached with NF as the driving force of pressure difference is finally equalized by the osmotic pressure of the feed. In addition, it can be seen that the curves of the three sub-trials fit very well together and draw a plausible overall curve of the permeate flux in dependency of increasing lactose concentration. This also means that no fouling layer has formed in the course of concentration [11], since the flux at the end of one sub-trial corresponds to the one at the beginning of the following sub-trial. So it can be concluded that the driving forces that determine the flux curve are mainly the increasing osmotic pressure and the concentration polarization effects. The fact that the flux decreases to a similar extent for all five membranes also is an indication that the increasing osmotic pressure and concentration polarization due to increasing lactose concentration are primarily responsible. Towards the end of the concentration, when a lactose concentration of approximately 170 g/L has been reached, the flux is reduced by around 80% compared to the initial value at 25 g/L, with only very small differences of around 79% for the SR3D and a maximum of 83% for the NFG.

Furthermore, the retention curve of the NFG membrane indicates that the higher lactose concentration on the membrane surface due to concentration polarization leads to an improved retention. Since the concentration ratios build up by concentration polarization are reset at the beginning of each sub-trial, and thus a higher part of the lactose permeates at the beginning of each trial, the NFG does not show a consistently stable retention throughout the three sub-trials. In contrast, the other membranes provide a constantly high lactose rejection of ≥93%, which tends to improve by an average of 2% with increasing degree of concentration.

As an intermediate conclusion, it can be stated that the NFG is not suitable due to the poor lactose retention. The other four membranes all have similarly high lactose retentions, so that the decision which membranes to be used for the scale-up was based on the permeate fluxes. Since the TS40 had the lowest flows, it was not considered any further. The SR3D would be a very interesting candidate, but KOCH Membrane Systems currently does not offer 1812″ spiral wound membranes, so that tests on a semi-pilot scale were not feasible and this circumstance points out another challenge researcher are faced with when membrane screening trials are planned. In conclusion, the scale-up was performed with the NFS and XN45, also because these two membranes differ in their MWCO which could have a varying outcome during the concentration of whey permeate.

3.3. Scale-Up and Performance Using Industrial Feeds

Two issues were the focus of the scale-up: First, comparison and evaluation of the results gained with the flat-sheet screening module with those of an 1812″ spiral wound module. Second, it should be evaluated to what extent the data obtained using the lactose model system is in agreement with the data when whey permeate is concentrated, because the whey permeate contains a large number of different components in addition to lactose as stated earlier, which influence the three above mentioned thermodynamic driving forces. Moreover, since in literature it is also described that NF membranes partially retain ions [11,12,14,15], this point was also evaluated using ion quantification and conductivity measurements.

3.3.1. Comparison of Module Performance

In Figure 5, the results of the concentration of the lactose model system using the flat-sheet membrane are compared with those using the spiral wound module. As expected for both membranes, the permeate flow rate through the spiral wound membrane is lower than through the flat-sheet membrane. Räsänen et al. [12] also compared four commercial NF membrane performances with flat-sheet trials and spiral wound modules and made comparable observations. The reason might be due to the different prevailing cross-flow conditions in the respective modules, different thickness of boundary layers, as the spiral wound module might offer a bigger resistance to the feed flow due to its multi-layered structure and the integrated spacers amplifying the effects of concentration polarization. However, van Gauwbergen et al. stated that geometry of the channel has little impact compared to the dominating effects of the spacer on the hydrodynamics [37]. It was shown that existing dead zones in spiral wound modules enhance local concentration polarization and result in a reduction of the overall membrane performance [37]. When the percentage drop in permeate flow over the course of the concentration at similar lactose concentrations is compared, the NFS shows a reduction of approximately 82% which is in very good agreement with that in the flat-sheet tests. The rejection values correspond equally well. In contrast, the use of the spiral wound module of the XN45 resulted in an approximately 11% reduced retention.

These results underline the need for scale-up experiments with drastically increased membrane areas, use of spacer and thus different hydrodynamics and/or modules closed to the ones used in industry, since filtration experiments with flat-sheet membranes have some potential weaknesses [12,38,39] as described in more detail below. Although [12] observed comparable retentions for flat-sheet as well as spiral wound modules, Schipolowski et al. [38] discussed and summarized critically the shortcomings of flat-sheet trials that are mainly due to a small effective membrane area and differences in test and flow conditions. Additionally, a certain heterogeneity in the membrane material and active layer as a result of the production process of the membrane cannot be avoided so that membrane manufacturers usually state the permeability of the membrane within a range of ±15–20% around the nominal value [20,38]. This fact might be one of the reasons why for NF membranes ranges for the MWCO are given, because the molecules and compounds that are to be separated with NF membranes are relatively small and therefore smaller fluctuations in the membrane structure might already have an influence on the membrane performance in terms of permeate flow and retention. When the available filtration area of flat-sheet membranes is limited, differences within the membrane structure can have a higher impact on the results, compared to spiral wound modules where the heterogeneity of the membrane structure is averaged over a larger area. However, Schipolowski et al. stated, that an increase in the filtration area through larger test cells does not necessarily alter the reliability since the production conditions for a nearby membrane area tends to produce a rather homogeneous membrane structure [38]. In the test setup applied in this study, the spiral wound module has an area that is almost 38 times larger than the tested flat-sheets. Taking into account the stated MWCO of the XN45 with 300–500 Da, the determined retentions of the 1812″ spiral wound module seem quite plausible and suggest that the flat-sheet membrane batch covered presumably the lower MWCO range and therefore gave a higher lactose retention. As seen in the parameter studies for the NFG as an example for a looser membrane and in accordance with the literature [31,32], temperature can have a distinct effect on retention. Although this was not seen in the parameter studies for the XN45 may be due to the tighter membrane structure of the flat-sheet batch, it was seen with the spiral wound module. Therefore, for the XN45 the concentration using the spiral wound module was repeated at 20 °C. As a result, the retention was improved by almost 9%, so that these adapted operating parameters were used for the subsequent concentration of the whey permeate. As already discussed, this led to a reduced permeate flow. At both temperatures, 20 °C and 45 °C, the flux decline with the XN45 was less pronounced during the concentration than with the NFS. Precisely at 45 °C by about 71% and at 20 °C by approximately 73% although the reduction of temperature went along with an improvement in retention. It is assumed that this could be a result of the slight lactose permeation that might affect the concentration polarization layer on the membrane surface, although this was not observed with the NFG in the parameter studies, in which even more lactose passed into the permeate during the course of the concentration. On the other hand, the initial permeate flux curve in the course of concentration for the NFG was relatively steep, which makes comparability difficult.

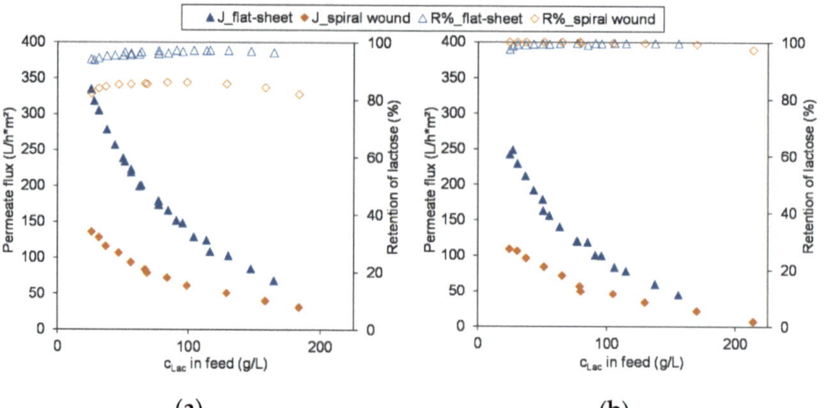

Figure 5. Comparison of membrane performance (flux (J) and retention (R%) of lactose (Lac)) of flat-sheet membrane versus 1812″ spiral wound module for concentration of lactose model solution; (**a**) XN45 (45 °C & 3 MPa); (**b**) NFS (45 °C & 2 MPa). Operation parameters summarized in Table 3.

Overall, after comparing the membrane performance of flat-sheet membranes and spiral wound membranes, it can be stated that flat-sheet membrane tests are well suited for preliminary investigations and a pre-selection, and in particular for a relative comparison of different membranes within the same test setup. Especially because in literature there are only scattered results with one [3,11,15] or up to four membranes [12]. However, it is usually difficult or impossible to compare the self-acquired data with those in the literature, since the experimental setup shows distinct differences for instance due to different test cells, plant construction and operating parameters. Therefore it was the aim of this study to give a relative comparison of ten membranes using the same test conditions. To finally be able to give reliable statements about the membrane performance, it is necessary to carry out tests with a larger filtration area, i.e., with spiral wound modules, since the production-related heterogeneity of the polymers will be statistically balanced with a larger filtration area. In particular, if the differences between the membrane performances of the flat-sheet membranes are only small, an adjustment using spiral wound modules is recommended. The "gold standard" would be using those modules and sizes that will be applied in the later industrial process, but as this is often not feasible, it is a good compromise for an approximation to use smaller module sizes where at least the flow conditions and retention might be comparable [38].

3.3.2. Concentration of Industrial Whey Permeate

Another objective of this study was to verify to which extent the selected approach, initially carrying out the membrane screening with a lactose model system instead of whey permeate, is advisable and appropriate and where there might be potential limitations.

In Figure 6 the permeate flux and the retention in the course of concentration of whey permeate are compared to that of the lactose model solution. For both membranes the lactose rejection is in very good agreement with the model system. The basic course of the permeate flux curves also agree well with the model system. Furthermore, the curves progress evenly, regardless of the subdivision into two sub-trials. This indicates that the remained proteins and peptides, which are still present in small amounts in the whey permeate (Table 1) because they cannot be retained by the 10 kDa UF membrane, do not build up a pronounced gel or fouling layer that would lead to a reduction in the flux as this was observed by [15] during the NF of sweet whey. However, in a long term process the formation of a gel layer might occur due a slowly but continuous accumulation of proteins and peptides which impairs the membrane performance. Therefore, it is recommended to conduct experiments for a much longer period of time with industrial feed in order to evaluate the long-term performance of the membrane and the process.

The fact that the permeate flux curves of the whey permeate and the lactose model system are not congruent is probably due to the high ion content of the whey permeate that affect the osmotic pressure of the feed. Timkin and Lazarev [40] have experimentally determined the osmotic pressure of a milk permeate and a lactose solution. Although the amount of lactose (4.5%) was much higher than the amount of ions (0.62%) the osmotic pressure of the milk permeate was approximately double as high as of the lactose solution indicating a pronounced effect of the milk ions on the osmotic pressure [40]. Since the osmotic pressure of the feed reduces the driving force of the process, this leads to a reduced permeate flow.

As shown in Figure 7, the conductivity increases in the course of concentration both in the retentate and in the permeate, with the conductivity in the retentate being significantly higher than that of the permeate right from the start. This means that in addition to the lactose, some of the ions are also retained and concentrated which was also confirmed by ion chromatography performed.

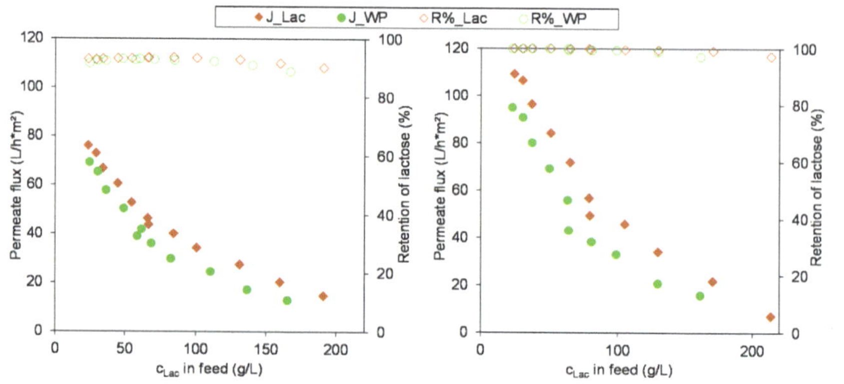

(a) (b)

Figure 6. Resulting permeate flux (J) and retention (R%) of lactose during concentration of lactose model solution (Lac) versus whey permeate (WP); (**a**) XN45 (20 °C & 3 MPa); (**b**) NFS (45 °C & 2 MPa). Operation parameters summarized in Table 3.

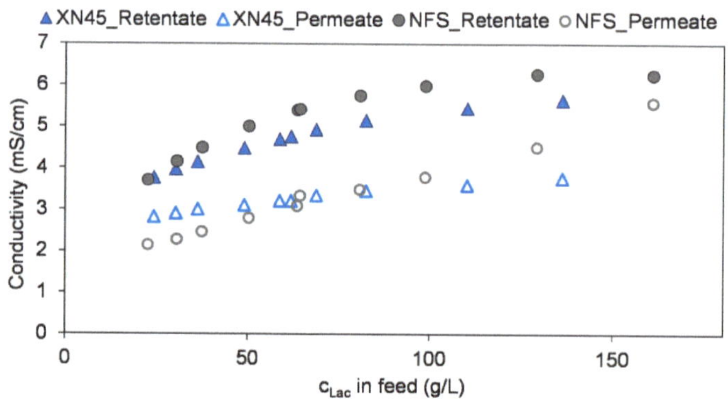

Figure 7. Comparative comparison of the course of conductivity in retentate and permeate during the concentration of lactose in whey permeate with XN45 and NFS membrane.

In Figure 8 the retentions of the analyzed ions are visualized. In accordance with literature, the divalent ions are almost completely (XN45) or completely (NFS) retained, while monovalent ions partially pass through the membrane. A special case are the chloride ions that show a negative retention due to the well-known Donnan effect. As the divalent anions cannot pass the membrane and thus the electrochemical potential increases on the feed side during the concentration, the monovalent chloride ions, as they are able to pass the membrane, are especially forced to permeate in order to equalize this state aiming towards an electro neutrality, even against its concentration gradient [11,14,15]. Finally, the observed chloride concentration in the permeate is higher than in the feed. Additionally, the increase in conductivity of the permeate and the course of retentions of ions demonstrate that increasing salt concentrations lead to a decrease in the retention of the ions due to electrostatic interactions [14,15,26]. Both in Figures 7 and 8 can be seen that the ion retention is less pronounced with the XN45 and thus a higher percentage of ions passes through the membrane. This results in a stronger partial demineralization in the course of lactose concentration compared to the NFS.

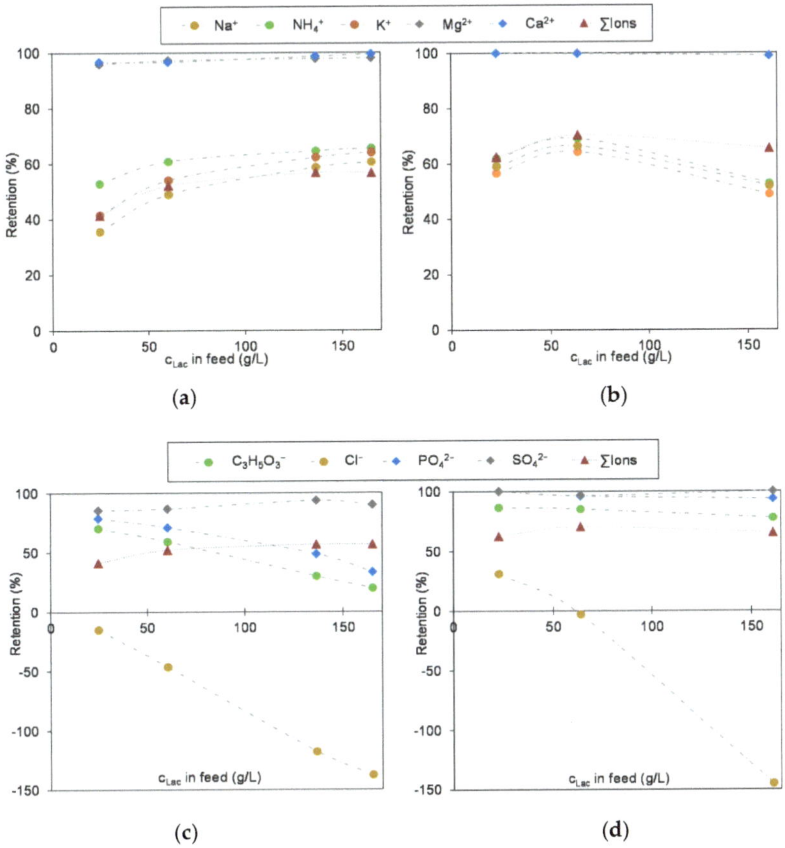

Figure 8. Retention of ions during concentration of lactose in whey permeate; (**a**) cations, membrane: XN45; (**b**) cations, membrane: NFS; (**c**) anions, membrane: XN45; (**d**) anions, membrane: NFS. Operation parameters summarized in Table 3.

Precisely, looking at the removal of single ions in dependency of the used membrane the following picture is drawn—lactic acid: 51% with the XN45 versus 20% with the NFS; chloride: 93% versus 70%; sodium: 42% versus 26% and potassium: 38% versus 24%. In turn this could explain why the permeate flux curves of the lactose and whey permeate feeds using the XN45 are in better agreement than the curves of the NFS. Since the soluble ingredients, especially lactose and ions, determine the osmotic pressure of the feed [19], the deviations between the curves increase in the course of concentration. Overall, it can be concluded that the course of the flux as the lactose is concentrated is basically a function of the concentration polarization effects, which in turn are determined by the osmotic pressure [11,35,40]. Thus, the lactose model solution can give a good approximation of the general membrane performance in terms of permeate flow and retention. However, since the raw material milk and its specific ionic composition are subject to a certain range of fluctuation as discussed earlier, trials with industrial whey feeds are necessary for a final evaluation and set-up of the whole process.

Another factor that might be relevant for the design of the process is the composition of the resulting permeate. Especially if for instance legislative specifications regarding the waste water treatment have to be obeyed or if the permeate is to be used as rinsing water or for DF in order to economize resources. Figure 9 shows the increase in lactose concentration in the retentate (y-primary axis) and permeate (y-secondary axis) with raising VCR for both feeds. Additionally, the figure visualizes the lactose concentration ratios, i.e., how they prevail directly at the membrane (permeate at the specific VCR) and how it emerges within the total cumulative permeate (P_cumulative). With the XN45, a small portion of the lactose permeates continuously, whereby this increases noticeably with progressive lactose concentration in the retentate, but is has a less pronounced influence on the cumulative permeate. When lactose of the model solution was concentrated to approximately 160 g/L, around 6.2% of lactose passed into the permeate. For whey permeate as feed, the ions appear to increase the lactose loss due to their concentration polarization enhancing effect, which is then about 10.4%. In contrast, the NFS exhibits complete retention over a longer period of time, up to around 80 g/L lactose within the feed. With rising concentrations, only minimal permeation of lactose takes place, so that for both feeds, the lactose model system (0.3%) and the whey permeate (0.7%), less than 1% lactose passes through the membrane and thus the resulting permeate is almost lactose-free and could e.g., be used for pre-cleaning, DF or other processes within the dairy.

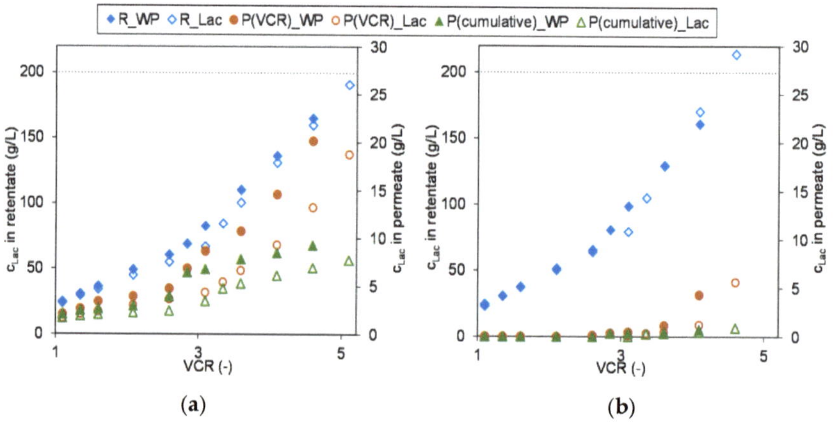

Figure 9. Evolution of lactose concentration in retentate (R), permeate (P(VCR)) and cumulative permeate (P(cumulative)) during the concentration of lactose model system (Lac) and whey permeate (WP); (**a**) membrane: XN45; (**b**) membrane: NFS.

4. Conclusions

The issue of identifying suitable membranes for an ecologically and economically efficient process of lactose concentration from whey permeate was realized in three stages including scale-up. First, 10 polymer membranes as flat-sheets were examined for their suitability within a TMP-screening, whereas a high to complete lactose retention in combination with a high flux were defined as the decision criterion. These trials were carried out using a lactose model solution, as whey permeate composition can vary widely, which could affect the results. Minimal coherence was found between the MWCO specified by the manufacturers and the maximum achievable flux underlining the necessity of preliminary trials. The retention of lactose was basically in agreement with the expectation based on the MWCO. In a second step detailed parameter studies were carried out with five promising selected membranes in order to investigate the influence and magnitude of the process variables temperature, TMP and increasing lactose content. Both a TMP increase from 2.0 to 3.0 MPa and a temperature increase from 20 °C to 45 °C led to significantly higher permeate fluxes, with the temperature influence being more pronounced and the combination of both operating parameters achieving the strongest effects. However, some membranes are limited in terms of their maximum tolerable process conditions, so that the synergistic effects of temperature and TMP cannot be used for them. With increasing lactose concentrations in the course of the concentration, there was a steady decrease in flux due to the increase in osmotic pressure and concentration polarization. In the last scale-up stage the plausibility and reproducibility of the results gained with the flat-sheet membranes were verified with spiral wound modules since flat-sheet results might be limited due to the small filtration area and diverse flow conditions compared to spiral wound modules used in industry. In addition, trials were carried out with real, industrially produced whey permeate. It was shown that the permeate flux values with the spiral wound modules were generally lower than with the flat sheets. With the NFS, the lactose retention was in good agreement with the preliminary tests. For the XN45, the retention was lowered with the spiral wound module at 45 °C, whereas a temperature reduction to 20 °C was accompanied by improved retention. The comparison of the membrane performance for the different feeds revealed that the lactose model solution is very well suited for preliminary studies in order to reproducibly depict the basic behavior of whey permeate. With increasing volume concentration ratio, there are more pronounced deviations between the flux curves of the model system and the whey permeate, because ions are also partially retained, which contribute to the amplification of the concentration polarization phenomena and the osmotic pressure of the concentrate. The more ions were retained, the more pronounced the differences were. The choice of membranes should be made depending on the objective which means whether the focus is more on complete lactose retention or partial demineralization. It could be shown that both membranes are basically well suited for concentration of lactose up to about 200 g/L. With the NFS, lactose can be concentrated with a loss of less than 1%, so that the resulting permeate is almost lactose-free and can be used for instance for upstream diafiltration steps or for cleaning. In contrast, a higher degree of demineralization can be achieved with the XN45, which is associated with higher lactose losses. Finally, in order to implement an ecologically and economically efficient process it is recommended to carry out trials with (a) industrial whey permeate and (b) performing long-time experiments, whereby the here presented data might provide a good starting point.

Author Contributions: Conceptualization, K.H. and C.H.; methodology, K.H. and C.H.; validation, K.H.; formal analysis, K.H.; investigation, K.H.; resources, C.H.; data curation, K.H.; writing—original draft, K.H.; writing—review and editing, C.H.; visualization, K.H.; supervision, C.H.; project administration, C.H.; funding acquisition, C.H. All authors have read and agreed to the published version of the manuscript.

Funding: This research was funded by the German Federal Ministry of Education and Research (BMBF) within the framework of Research at Universities of Applied Sciences 2018, grant number 13FH574IX6.

Institutional Review Board Statement: Not applicable.

Data Availability Statement: Not applicable.

Acknowledgments: The authors gratefully acknowledge the funding by the German Federal Ministry of Education and Research. Furthermore, the authors would like to appreciate the project partner Bayerische Milchindustrie eG (BMI) for providing industrial whey permeate. The authors gratefully thank the students Michel Ginzel, Johannes Jeske and Toni Schurzmann who performed part of the experiments during the course of their studies. Moreover, the authors acknowledge support by the German Research Foundation (DFG)-project number 491460386—and the Open Access Publishing Fund of Anhalt University of Applied Sciences.

Conflicts of Interest: The authors declare that they have no known competing financial interests or personal relationships that could have appeared to influence the work reported in this paper.

References

1. De Souza, R.R.; Bergamasco, R.; da Costa, S.C.; Feng, X.; Faria, S.H.B.; Gimenes, M.L. Recovery and purification of lactose from whey. *Chem. Eng. Process.* **2010**, *49*, 1137–1143. [CrossRef]
2. Mawson, A.J. Bioconversions for whey utilization and waste abatement. *Bioresour. Technol.* **1994**, *47*, 195–203. [CrossRef]
3. Atra, R.; Vatai, G.; Bekassy-Molnar, E.; Balint, A. Investigation of ultra- and nanofiltration for utilization of whey protein and lactose. *J. Food Eng.* **2005**, *67*, 325–332. [CrossRef]
4. Gänzle, M.G.; Haase, G.; Jelen, P. Lactose: Crystallization, hydrolysis and value-added derivatives. *Int. Dairy J.* **2008**, *18*, 685–694. [CrossRef]

5. Smithers, G.W. Whey and whey proteins—From 'gutter-to-gold'. *Int. Dairy J.* **2008**, *18*, 695–704. [CrossRef]
6. Silanikove, N.; Leitner, G.; Merin, U. The Interrelationships between Lactose Intolerance and the Modern Dairy Industry: Global Perspectives in Evolutional and Historical Backgrounds. *Nutrients* **2015**, *7*, 7312–7331. [CrossRef]
7. Skoet, J.; Gerosa, S. *Milk Availability: Trends in Production and Demand and Medium-Term Outlook*; Food and Agriculture Organization: Rome, Italy, 2012.
8. Mueller, I.; Kiedorf, G.; Runne, E.; Pottratz, I.; Seidel-Morgenstern, A.; Hamel, C. Process Control and Yield Enhancement of the Galacto-Oligosaccharide Formation. *Chem. Ing. Tech.* **2018**, *90*, 725–730. [CrossRef]
9. Pottratz, I.; Mueller, I.; Hamel, C. Potential and Scale-Up of Pore-Through-Flow Membrane Reactors for the Production of Prebiotic Galacto-Oligosaccharides with Immobilized β-Galactosidase. *Catalysts* **2022**, *12*, 7. [CrossRef]
10. Pázmándi, M.; Maráz, A.; Ladányi, M.; Kovács, Z. The impact of membrane pretreatment on the enzymatic production of whey-derived galacto-oligosaccharides. *J. Food Process Eng.* **2018**, *41*, e12649. [CrossRef]
11. Cuartas-Uribe, B.; Alcaina-Miranda, M.I.; Soriano-Costa, E.; Mendoza-Roca, J.A.; Iborra-Clar, M.I.; Lora-García, J. A study of the separation of lactose from whey ultrafiltration permeate using nanofiltration. *Desalination* **2009**, *241*, 244–255. [CrossRef]
12. Räsänen, E.; Nyström, M.; Sahlstein, J.; Tossavainen, O. Comparison of commercial membranes in nanofiltration of sweet whey. *Lait* **2002**, *82*, 343–356. [CrossRef]
13. Zadow, J.G. *Whey and Lactose Processing*; Springer: Dordrecht, The Netherlands, 1992; pp. 2–13, ISBN 9789401128940.
14. Melin, T.; Rautenbach, R. *Membranverfahren: Grundlagen der Modul- und Anlagenauslegung; mit 76 Tabellen, 3*; Aktualisierte und erw. Aufl.; Springer: Berlin, Germany, 2007; pp. 285–306, ISBN 3-540-00071-2.
15. Suárez, E.; Lobo, A.; Álvarez, S.; Riera, F.A.; Álvarez, R. Partial demineralization of whey and milk ultrafiltration permeate by nanofiltration at pilot-plant scale. *Desalination* **2006**, *198*, 274–281. [CrossRef]
16. Qi, T.; Yang, D.; Chen, X.; Qiu, M.; Fan, Y. Rapid removal of lactose for low-lactose milk by ceramic membranes. *Sep. Purif. Technol.* **2022**, *289*, 120601. [CrossRef]
17. Straatsma, J.; Bargeman, G.; van der Horst, H.C.; Wesselingh, J.A. Can nanofiltration be fully predicted by a model? *J. Membr. Sci.* **2002**, *198*, 273–284. [CrossRef]
18. Hiddink, J.; de Boer, R.; Nooy, P. Reverse Osmosis of Dairy Liquids. *J. Dairy Sci.* **1980**, *63*, 204–214. [CrossRef]
19. Mason, E.A.; Lonsdale, H.K. Statistical-mechanical theory of membrane transport. *J. Membr. Sci.* **1990**, *51*, 1–81. [CrossRef]
20. Petersen, R.J. Composite reverse osmosis and nanofiltration membranes. *J. Membr. Sci.* **1993**, *83*, 81–150. [CrossRef]
21. Chen, B.-Z.; Ju, X.; Liu, N.; Chu, C.-H.; Lu, J.-P.; Wang, C.; Sun, S.-P. Pilot-scale fabrication of nanofiltration membranes and spiral-wound modules. *Chem. Eng. Res. Des.* **2020**, *160*, 395–404. [CrossRef]
22. Tsuru, T.; Izumi, S.; Yoshioka, T.; Asaeda, M. Temperature effect on transport performance by inorganic nanofiltration membranes. *AIChE J.* **2000**, *46*, 565–574. [CrossRef]
23. Nilsson, M.; Trägårdh, G.; Östergren, K. The influence of pH, salt and temperature on nanofiltration performance. *J. Membr. Sci.* **2008**, *312*, 97–106. [CrossRef]
24. Xiao, H.-F.; Shao, D.-D.; Wu, Z.-L.; Peng, W.-B.; Akram, A.; Wang, Z.-Y.; Zheng, L.-J.; Xing, W.; Sun, S.-P. Zero liquid discharge hybrid membrane process for separation and recovery of ions with equivalent and similar molecular weights. *Desalination* **2020**, *482*, 114387. [CrossRef]
25. Wijmans, J.G.; Baker, R.W. The solution-diffusion model: A review. *J. Membr. Sci.* **1995**, *107*, 1–21. [CrossRef]
26. Hagmeyer, G.; Gimbel, R. Modelling the salt rejection of nanofiltration membranes for ternary ion mixtures and for single salts at different pH values. *Desalination* **1998**, *117*, 247–256. [CrossRef]
27. Nilsson, M.; Trägårdh, G.; Östergren, K. The influence of sodium chloride on mass transfer in a polyamide nanofiltration membrane at elevated temperatures. *J. Membr. Sci.* **2006**, *280*, 928–936. [CrossRef]
28. Bargeman, G.; Vollenbroek, J.M.; Straatsma, J.; Schroën, C.; Boom, R.M. Nanofiltration of multi-component feeds. Interactions between neutral and charged components and their effect on retention. *J. Membr. Sci.* **2005**, *247*, 11–20. [CrossRef]
29. Koros, W.J.; Ma, Y.H.; Shimidzu, T. Terminology for membranes and membrane processes (IUPAC Recommendations 1996). *Pure Appl. Chem.* **1996**, *68*, 1479–1489. [CrossRef]
30. Dutré, B.; Trägardh, G. Macrosolute-microsolute separation by ultrafiltration: A review of diafiltration processes and applications. *Desalination* **1994**, *95*, 227–267. [CrossRef]
31. Yao, W.X.; Kennedy, K.J.; Tam, C.M.; Hazlett, J.D. Pre-treatment of kraft pulp bleach plant effluent by selected ultrafiltration membranes. *Can. J. Chem. Eng.* **1994**, *72*, 991–999. [CrossRef]
32. Ben Amar, N.; Saidani, H.; Deratani, A.; Palmeri, J. Effect of temperature on the transport of water and neutral solutes across nanofiltration membranes. *Langmuir ACS J. Surf. Colloids* **2007**, *23*, 2937–2952. [CrossRef]
33. Mänttäri, M.; Pihlajamäki, A.; Kaipainen, E.; Nyström, M. Effect of temperature and membrane pre-treatment by pressure on the filtration properties of nanofiltration membranes. *Desalination* **2002**, *145*, 81–86. [CrossRef]
34. Schmidt, C.M.; Mailänder, L.K.; Hinrichs, J. Fractionation of mono- and disaccharides via nanofiltration: Influence of pressure, temperature and concentration. *Sep. Purif. Technol.* **2019**, *211*, 571–577. [CrossRef]
35. Wijmans, J.G.; Nakao, S.; Smolders, C.A. Flux limitation in ultrafiltration: Osmotic pressure model and gel layer model. *J. Membr. Sci.* **1984**, *20*, 115–124. [CrossRef]
36. Di Giacomo, G.; Scimia, F.; Taglieri, L. Cost-effective disposal of milk whey II: Recovery and purification of lactose and pure water from the diafiltration permeate stream. *DWT* **2017**, *76*, 339–342. [CrossRef]

37. Van Gauwbergen, D.; Baeyens, J. Macroscopic fluid flow conditions in spiral-wound membrane elements. *Desalination* **1997**, *110*, 287–299. [CrossRef]
38. Schipolowski, T.; Jeżowska, A.; Wozny, G. Reliability of membrane test cell measurements. *Desalination* **2006**, *189*, 71–80. [CrossRef]
39. Jeżowska, A.; Schipolowski, T.; Wozny, G. Reduction of random and systematic errors of test cell measurements. *Desalination* **2006**, *200*, 160–162. [CrossRef]
40. Timkin, V.A.; Lazarev, V.A. Determination of the osmotic pressure of multicomponent solutions in the food industry. *Pet. Chem.* **2015**, *55*, 301–307. [CrossRef]

membranes

Article

The Performance of Ultrafiltration Process to Further Refine Lactic Acid from the Pre-Microfiltered Broth of Kitchen Waste Fermentation

Yan Guo [1], Chenglong Li [1], Hongjun Zhao [1], Xiaona Wang [1,*], Ming Gao [1], Xiaohong Sun [2,*] and Qunhui Wang [1]

[1] Department of Environmental Engineering, University of Science and Technology Beijing, 30 Xueyuan Road, Haidian District, Beijing 100083, China

[2] Beijing Agro-Biotechnology Research Center, Beijing Academy of Agriculture and Forestry Sciences, Beijing 100097, China

* Correspondence: wangxiaona071@163.com (X.W.); sunxiaohong19675@aliyun.com (X.S.)

Abstract: Lactic acid (LA) is an important chemical material facing rapid demand in recent years. The oriented fermentation of kitchen waste is a promising route for economic LA production. However, the refinement of LA from fermentation broth is a spiny issue. In this study, the performance of ultrafiltration (UF) process for the refinement of LA from the pre-microfiltered broth of kitchen waste fermentation was first investigated. The results showed that with 50 KDa polyethersulfone membrane, under the optimum pressure of 120 KPa, the pH of 6.0, and the backflushing mode with the deionized water for 3 min, the best performance was achieved with the chroma removal efficiency, turbidity removal efficiency, protein removal efficiency and total sugar removal efficiency of 54.3%, 89.8%, 71.7% and 58.5%, respectively, and LA recovery efficiency was 93.6%. The results indicated that the UF process could further effectively refine the pre-microfiltered broth of kitchen waste fermentation, and the combination of microfiltration and UF process is ideal for achieving desirable LA refinement performance. This study verified the feasibility of UF process in LA refinement from pre-microfiltered broth of kitchen waste fermentation, and based on the results, the further exploration of proper post-process to treat UF filtrate for obtaining LA product with higher quality should be explored in the future.

Keywords: kitchen waste; fermentation broth; lactic acid; ultrafiltration; refinement; flushing

Citation: Guo, Y.; Li, C.; Zhao, H.; Wang, X.; Gao, M.; Sun, X.; Wang, Q. The Performance of Ultrafiltration Process to Further Refine Lactic Acid from the Pre-Microfiltered Broth of Kitchen Waste Fermentation. *Membranes* **2023**, *13*, 330. https://doi.org/10.3390/membranes13030330

Academic Editor: Daqi Cao

Received: 10 January 2023
Revised: 19 February 2023
Accepted: 9 March 2023
Published: 13 March 2023

1. Introduction

Lactic acid (LA) is one of the organic acids. The chemical formula of the LA is $C_3H_6O_3$. It has two optical isomers, Levo and Dextro, making itself a chiral molecule. L-isomers are commonly present among living organisms. The extended formula of LA is $CH_3CH(OH)CO_2H$, and it has a molar mass of 90.08 g/mol. Currently, as one of the important chemical materials for industrial applications, such as those related to the food, tanning, petrochemicals, pharmaceuticals, cosmetics and chemicals industry, LA is obtained mainly relying on chemical conversion from petrochemical raw material [1]. With the world's most urgent mission-carbon neutrality by the middle of the century, as well as the raised awareness of environmental effects in terms of pollution and greenhouse effect, the alternative route for LA production is expected [2,3]. Fortunately, it was reported that LA can be obtained through low-cost fermentation/anaerobic digestion process with the low-cost biomass [4,5]. However, the refinement and consequent concentration of LA from fermented broths are challenging due to the relatively low concentration of LA and complex impurities in the broths [6]. Thus, low cost and concise refinement technologies are urgently needed [7].

Research progress in the LA refinement was remarkable recently, together with the detection methods [8,9], and refinement techniques [6,10]. Various methods for refining LA from fermented broths were explored, including electrodialysis [11], chemical precipitation [12], ion-exchange [13], solvent extraction [14], distillation [15,16], adsorption [17] and membrane processes [6,18]. Especially, the membrane-based technology, characterized as being potentially efficient, cost-effective and eco-friendly, is a promising process [19].

Until now, the refinement of LA existing in the fermented broth was successfully conducted by the nanofiltration (NF) and reverse osmosis (RO) processes [20,21]. However, the pre-treatment of fermented broth is requested due to the high turbidity of broth, which cannot be directly fed to the spiral-wound membrane modules of the NF and RO processes. Commonly, the microfiltration (MF) and ultrafiltration (UF) processes, deemed as the best pre-treatment strategies for removing most components (such as particles, colloids, bacteria and high molecular mass organic matter) causing the membranes fouling [21], can effectively achieve this goal [20,22,23].

Food waste (FW), the main composition of municipal solid waste (MSW), accounts for 37–62% of MSW with a rapid increase trend in China and approximately 12–30% in some developed countries such as USA and UK [24]. Thus, FW is a promising source for energy and resource recovery, due to its high organic content [25]. In recent years, fermentation technology was adopted to recycle the FW through the production of many high-value products such as LA and alcohols existing in the fermented broths [26]. It seemed that the membrane process could be utilized to refine the fermented broth to obtain the high-value-added LA product. However, a concern was that the impact of fermented broths on membrane fouling is several times larger compared to that of the surface water and seawater [27].

In the author's previous lab-scale research, the MF filtration of the kitchen waste (KW) fermentation broth was conducted. After MF filtration, most of the suspended particles and bacteria in the broth were effectively removed. However, there are still many soluble macromolecular organic impurities in the MF permeate, such as protein substances and residual total sugars. Among them, due to the Maillard reaction, substances such as sugar and protein will produce brown or even black substances, which seriously affects the quality of LA products [28]. UF is a purification process that separates particulate matter from soluble compounds using an ultrafine membrane media. Ultrafiltration is an excellent separation technology for desalination pretreatment, reverse osmosis pretreatment, and wastewater reclamation, as well as for producing potable water. Like MF, UF is based on size exclusion or particle capture. UF can purify and concentrate macromolecular (10^3–10^6 Da) solutions, especially protein solutions. Therefore, it is desirable to use UF process to further refine the MF permeates to improve the LA purity.

Thus, in this study, the refinement of pre-microfiltered broth of kitchen waste fermentation was explored through the UF process. Firstly, the refinement performance of four different UF membranes on the simulated broth was compared for selecting the proper UF membrane type. Then, the selected UF membrane was adopted to refine the actual pre-microfiltered broth, and the effect of different operating pressures (0.08 MPa, 0.10 MPa, 0.12 MPa, 0.14 MPa) and different pH (5, 6, 7, 8) on refinement performance were investigated to optimize the operation parameters. Finally, the recovery efficiency of pure water flux of UF membrane under different cleaning modes, cleaning agents and cleaning time was conducted to obtain the optimal scheme for UF membrane cleaning. In addition, the 10 filtration-cleaning cycles were continuously carried out to investigate the effect of cleaning times on the treatment performance of UF membrane. Under each condition, the chroma removal efficiency (CRE), the turbidity removal efficiency (TRE), total sugar removal efficiency (SRE), protein removal efficiency (PRE) and LA recovery efficiency (LRE) were investigated.

2. Materials and Methods

2.1. *Fermented Broth*

2.1.1. The Pre-Microfiltered Broth

The broth was obtained through the fermentation of KW. KW was collected from the HongboYuan dining hall, located in the University of Science and Technology Beijing, China. The collected KW was manually homogenized by first sorting out the coarse impurities, such as the hard bones, plastic bags, paper towels and other impurities, and then placed into the meat grinder for mixing well. Then, the KW serum was transferred into the self-sealing bag, and stored at −20 °C freezer [29]. After thawing, tap water was added to the KW serum with the ratio of 1:1 (w/v) ratio to adjust the total solids concentration. The LA bacteria adopted were *Montessori Enterococcus* CGMCC 22227, with the inoculation ratio of 10%, the fermentation temperature of 43 °C and the pH of 6.8–7.0 controlled by NaOH solution (10 mol/L) every 12 h.

After the 84 h of fermentation, the fermentation remnant was centrifuged by centrifuge (Hunan Hexi Instrument Equipment Co., Ltd., Changsha, China) (speed 12,000 rpm, centrifugation 10 min). After centrifugation, the bottom sedimentary substance and the upper oily substance were abandoned to obtain the raw broth. The raw fermented broth had a pH of 6.2–6.3, an LA content of 63.96 g/L, a chroma of 1137 Hazen, a turbidity of 175 NTU, a protein content of 0.21 g/L and a total sugar content of 3.33 g/L. Finally, the raw fermented broth was pre-treated by the MF process under the condition of the optimum pressure of 100 KPa, the pH of 6.0, and the backflushing mode with the deionized water for 3 min, and the obtained MF filtrate was the pre-microfiltered broth in this research with an LA content of 58.78 g/L, a chroma of 445.7 Hazen, a turbidity of 9.8 NTU, a protein content of 0.08 g/L and a total sugar content of 2.14 g/L.

2.1.2. The Simulative Broth

To ensure the stability of the properties of the broth for comparison and selection of UF membranes, this study was equipped with simulative broth with reference to the previous research of our group. The simulative broth was prepared as follows: LA (47.9 g/L), bovine serum protein (2.5 g/L), glucose 3 g/L.

2.2. *Experimental Apparatus*

The UF apparatus adopted in this experiment was in a filtration cup configuration with a removable flat membrane structure (Shandong Bona Biotechnology Co., Ltd., Jinan, China), and the apparatus is shown in Figure 1. The magnetic stirring component was equipped to mix the liquid when conducting the filtration, and the nitrogen gas from the nitrogen gas cylinder (Beijing Huanyu Jinghui Gas Co., Ltd., Beijing, China) was used to force the fermented broth to pass through the membrane. The hydrophobic effect between the membrane surface and the proteins was deemed to be the dominant factor causing membrane fouling by the adsorption of proteins on the membrane. Thus, more hydrophilic membranes are usually favored [27]. In preliminary experiment, the MWCO range of under 150 KDa was proved to be reasonable for obtaining the desirable impurities, and so, based on the accessibility and cost, the 30 KDa, 50 and 100 KDa were selected. The four hydrophilic UF membranes were regenerated cellulose (RC) membranes (Shandong Bona Biotechnology Co., Ltd.) with a molecular weight cut off (MWCO) of 30 KDa (RC-30), 50 KDa (RC-50) and 100 KDa (RC-100), respectively, and polyethersulfone (PES) (Shandong Bona Biotechnology Co., Ltd.) membranes with a molecular weight of 50 KDa (PES-50) (In preliminary experiment, the 30 KDa PES and 100 KDa were also investigated. However, for the 30 KDa PES, the membrane flux was very low, and the blocking was serious. For the 100 KDa PES, the impurities removal efficiency was too poor. Thus, in this experiment, the 30 KDa PES and 100 KDa were excluded).

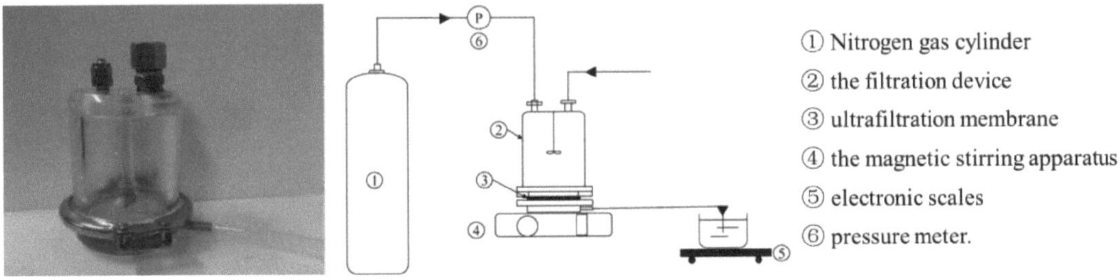

Figure 1. The illustration of the filtration configuration.

During the experiment, the UF membrane was set in the filtration cup, and then, a certain amount of fermented broth was loaded into the filtration cup, and the magnetic stirrer below was turned on at the speed of 600 rpm. Then, the nitrogen cylinder connected to the filtration cup was opened. By adjusting the pressure, the filtration was performed until the outflow velocity of the liquid was less than 0.2 mL/min.

When the UF membrane was not in use, it was stored in a container filled with desiccant. After use, it was kept in a wet state in 0.5% formaldehyde solution. It was soaked in deionized water for 1 h before each use.

2.3. Analysis Items and Methods

2.3.1. Turbidity and Chroma

The turbidity and chroma of the broth were analyzed by the water quality analyzer (Beijing Lianhua Yongxing Science and Technology Development Co., Ltd., Beijing, China). The TRE and the CRE are calculated, as shown in Equations (1) and (2) [28].

$$TRE = \frac{C_{n0} - C_n}{C_{n0}} \times 100 \tag{1}$$

where C_n is the turbidity of the filtered transmissible fluid, NTU; C_{n0} is the turbidity of the fermentation broth before filtration, NTU.

$$CRE = \frac{C_{h0} - C_h}{C_{h0}} \times 100 \tag{2}$$

where C_h is the chromaticity of the filtered liquid, Hazen; C_{h0} is the chromaticity of the fermentation broth before filtration, Hazen.

2.3.2. The Mass of LA

The determination of LA in broths was performed by high performance liquid chromatography (Shimadzu Corporation, Kyoto, Japan).

The samples were centrifuged in a 12,000 r/min for 10 min, and the supernatant was filtered by a 0.22 μm membrane before determination. The shodex Sugar SH1011 liquid chromatography columns (8.0 mm × 300 mm) and the RID detector were adopted. Chromatographic conditions were column temperature: 60 °C; mobile Phase: 5 mM H2SO4; flow rate: 1.0 mL/min; injection volume: 10 μL.

The calculation formula for the LRE is shown in Equation (3) [28].

$$LRE = \frac{C_L' V'}{C_{L0} V_0} \times 100 \tag{3}$$

where C_L' is the LA concentration of the fermented broth after filtration, g/L; V′ is the product of fermented liquid after filtration, L; C_{L0} is the LA concentration of fermentation broth before filtration, g/L; V_0 is the product of fermentation liquid before filtration, L.

2.3.3. The Mass of Proteins

Proteins were determined using the Coomassie Brilliant Blue G-250 method. The PRE is calculated, as shown in Equation (4) [28].

$$PRE = \frac{C_{p0} - C_p}{C_{p0}} \times 100 \quad (4)$$

where C_p is the protein concentration in the filtered solution, g/L; C_{p0} is the protein concentration in fermentation broth before filtration, g/L.

2.3.4. The Mass of Total Sugars

The determination of total sugars in the fermentation broth was carried out by the phenol-sulfuric acid method. The SRE is calculated, as shown in Equation (5) [28].

$$SRE = \frac{C_{s0} - C_s}{C_{s0}} \times 100 \quad (5)$$

where C_s is the total sugar concentration in the filtered permeable solution, g/L; C_{s0} is the total sugar concentration of fermentation broth before filtration, g/L.

2.3.5. Membrane Flux

Membrane flux was defined as the volume of liquid that permeates a membrane per unit area per time. The flux calculation formula is shown in Equation (6) [28].

$$Flux\left(\mathrm{mL/min} \times \mathrm{m}^2\right) = \frac{V}{t \times A} \times 100 \quad (6)$$

where V is the filtrate, mL; t is the filtration time, min; A is the membrane area, m^2.

2.3.6. The Membrane Cleaning Performance

The membrane cleaning performance was expressed by the membrane pure water flux recovery efficiency (FRE), which was calculated based on Equation (7) [28]

$$FRE = \frac{J_w - J_{fw}}{J_0 - J_{fw}} \times 100 \quad (7)$$

where J_0 is pure water permeation flux before membrane use, $L/m^2/h$; J_{fw} is pure water permeation flux after membrane contamination, $L/m^2/h$; J_w is pure water permeation flux after membrane cleaning, $L/m^2/h$.

3. Results

3.1. Comparison and Selection of UF Membranes

At the room temperature of about 25 °C, and under the operating pressure of 120 KPa, four kinds of UF membranes were used to treat the simulative broth separately. The SRE, PRE and LRE were compared to select the suitable UF membrane for the refinement of LA from the pre-microfiltered broth.

Under different pressures, the pure water flux of the four UF membranes changed, as shown in Figure 2a, and with the increase in the operating pressure, the pure water flux of the four membranes showed a linear upward trend. Under the same pressure, the pure water flux of the PES-50 membrane was significantly higher than that of the other three kinds of UF membranes, which indicated that the permeability of the PES-50 membrane was higher, and the membrane resistance was lower. It was explained by the hydrophobicity of RC membrane exceeding that of the PES membrane with the increase in pressure, as the higher the pressure, the more hydrophobic the membrane became [30]. When the simulative broth was filtered with four kinds of UF membranes, the performance of each membrane was shown in Figure 2b. The PREs were between 97.6% and 98.9%, which showed that the UF membrane had a very good removal performance for macromolecular proteins. Meanwhile, for the RC membrane, the PRE increased with the decrease in pore size, but the overall change range was very small, i.e., within 1%. Like protein, the SRE also increased with the decrease in pore size, but compared with the PRE, it was evident that the removal performance of UF membrane on total sugar was weak, i.e., between 5.9% and 37.0%. This was because the pore size determines the separation performance of UF: the UF process can effectively separate macromolecular substances (such as proteins), while most of the small molecular substances (such as glucose) will pass through the UF membrane [31,32]. For LA recovery, it was found that with the decrease in pore size, LRE showed a linear downward trend, and the decline was evident. For the PES-50 membrane, it was found that the SRE was 34.6% under the same MWCO, which was significantly higher than the RC-50 membrane (its SRE was only 12.2%). The RC-100 membrane had the lowest SRE, which indicated that the most sugar in the broth had lower MWCO. At the same time, the PES-50 membrane had the best PRE of 98.9%, while LRE of 90.1% was slightly lower than that of the RC-50 membrane (90.4%). Comprehensively, considering that the PES membrane had a high SRE and PRE, and the desirable LRE, PES-50 was finally selected for refining LA broth of KW fermentation.

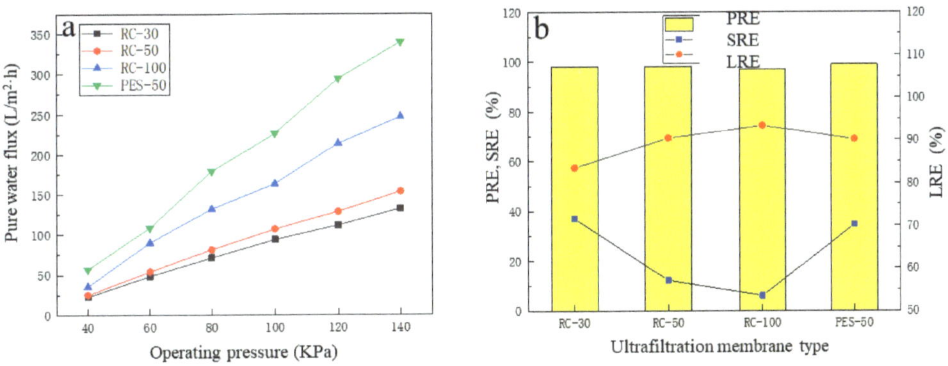

Figure 2. The selection of ultrafiltration membrane type: (**a**) the pure water flux of four kinds of ultrafiltration membrane under different operating pressure; (**b**) the filtration performance of four kinds of ultrafiltration membrane for protein, total sugar and lactic acid in simulated broth.

3.2. The Effect of Operating Pressure on LA Refinement Performance

After the PES-50 membrane was selected, the actual pre-microfiltered broth was tested. To investigate the effect of operating pressure on LA refinement performance, experiments were carried out under different operating pressures. The change of membrane flux over the time was recorded, and the TRE, CRE, PRE, SRE and LRE were analyzed.

3.2.1. The Change of UF Membrane Flux

As observed in Figure 3a, the higher the operating pressure, the higher the membrane flux. In initial time, the membrane flux decreased rapidly. For this phenomenon, the reason was inferred as at the beginning of filtration, the broth flowed to the surface of the membrane driven by the static pressure difference, and the macromolecular solute accumulated rapidly on the surface of the membrane due to the interception of the membrane. Thus, there existed concentration polarization phenomenon between the surface of the membrane and the broth, and the concentration polarization layer formed. The formation of concentration polarization made the resistance increase for broth reach the surface of the membrane. Thus, membrane flux decreased rapidly. With the filtration continued, the concentration polarization gradually reached a stable state, and the flux of the film maintained a stable state [33]. At 25–55 min, it was observed that the downward trend of membrane flux under four pressure conditions significantly reduced, and the membrane flux tended to be stable. When the operating pressure was below 120 KPa, the higher the operating pressure, the higher the membrane flux. However, when the operating pressure continued to increase to 140 KPa, the membrane flux basically did not increase, not even slightly less than the membrane flux under 120 KPa. This may have been due to the increased pressure caused concentration polarization and blockage of the membrane, then the formation of the cake layer caused a stable flux [34]. Additionally, the membrane flux under 120 KPa and 140 KPa was very close and significantly higher than that under 80 KPa and 100 KPa. After 55 min, the membrane fluxed under 120 KPa and 140 KPa, and began to plummet strictly again, while the membrane flux under 80 KPa and 100 KPa maintained a slow downward trend. It was probably because during the initial 55 min, with the high membrane flux under 120 KPa and 140 KPa, after most of the clear liquid having flowed out through the UF membrane, the concentrations of protein, total sugar and other substances in the remaining broth were significantly increased, which further aggravated the UF membrane pollution, and so, the membrane flux decreased rapidly [35]. Therefore, from the perspective of membrane flux, it was more appropriate to choose an operating pressure of 120 KPa.

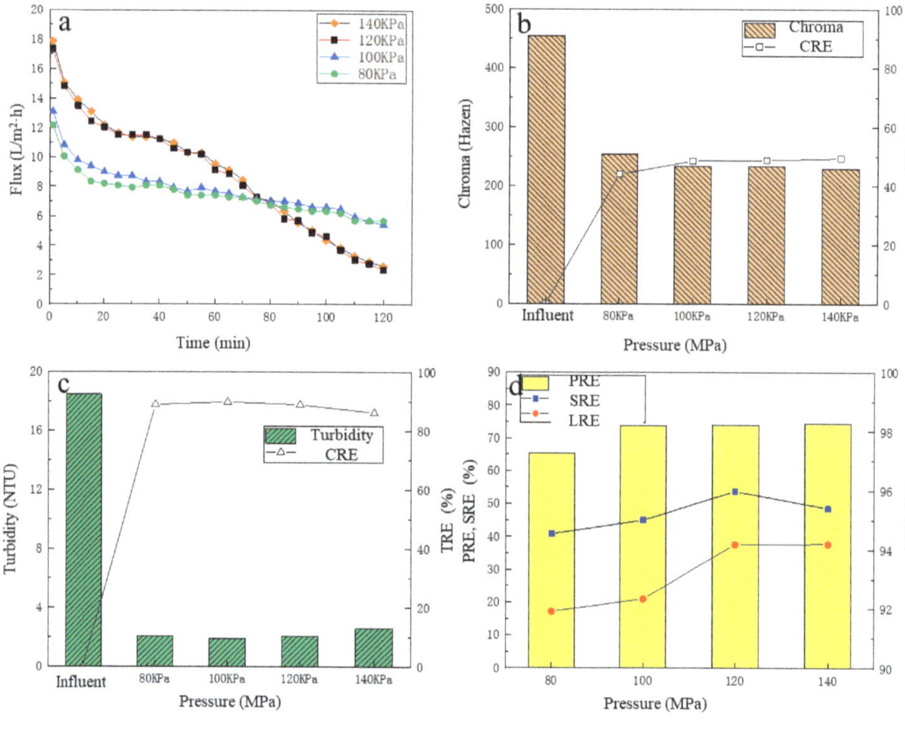

Figure 3. The effect of different pressure on filtration performance: (**a**) the flux change; (**b**) the chroma removal; (**c**) the turbidity removal; (**d**) the protein removal efficiency, the total sugar removal efficiency, and the lactic acid recovery efficiency.

3.2.2. The Removal of Chroma and Turbidity

Before and after filtration, the chroma changes of the pre-microfiltered broth are shown in Figure 3b. The chroma of the filtrate was between 230–250 Hazen. It was clear that the CRE increased with the increase in operating pressure, reaching a maximum value of 49.38% under 140 KPa. Overall, the CRE fluctuated slightly. The reason may be that some macromolecular pigments were removed by the UF membrane, but the remaining pigment molecules with a relatively lower molecular weight [36] passed through the UF membrane, so that the chroma of the filtrate was still relatively high.

Before and after filtration, the turbidity of the pre-microfiltered broth changed, as shown in Figure 3c. As stated in Section 2.1.1, most of the insoluble particulate matter in the raw fermentation broth was removed by the pre-MF process, and only a small part of the particles with smaller particle size remained in the pre-microfiltered broth, which was further removed by UF membrane. The turbidity of the filtrate was basically between 2.1 and 2.8 NTU, which was extensively lower than that of the feed. With the operating pressure increased, the TRE of the fermentation broth increased first and then decreased, but the overall change was small. Under 120 KPa, the turbidity of the filtrate was as low as 2.14 NTU with the TRE of 85.62%.

3.2.3. The Removal of Protein, Total Sugar and the LA Recovery

Under different operating pressures, the refinement performance of UF membrane for the pre-microfiltered broth is shown in Figure 3d. The PRE increased with the increase in the operating pressure. However, when the operating pressure was higher than 80 KPa, the increase in PRE was not obvious, i.e., around 74%. With the increase in operating pressure, the SRE was basically on the rise, reaching a maximum of 53.77% under 120 KPa. When the operating pressure further increased to 140 KPa, the SRE reduced to a certain extent, which was 48.62%. For LA recovery, up to 120 KPa, the LRE increased with the increase in the operating pressure. Under 120 KPa, the LRE was 94.19%. It was inferred that the higher the operating pressure, the larger the filtrate and the higher the LRE. When the operating pressure further increased to 140 KPa, the LRE remained basically unchanged. This was because when the pressure continued to increase, the membrane flux no longer increased, and the membrane surface contamination layer was tighter, finally hindering the passage of LA, resulting in an insignificant change in LA recovery. Considering the PRE, SRE and LRE, 120 KPa was finally selected as the optimal operating pressure for UF treatment of pre-microfiltered broth.

3.3. The Effect of pH on UF Performance

In addition to the operating pressure, the pH of the pre-microfiltered broth also had an impact on the UF process. To further investigate the effect of pH on UF performance, an HCl solution of 6 mol/L and a NaOH solution of 1 mol/L were used in this study to adjust the pH of the pre-microfiltered fermented broth to 5, 6, 7 and 8, respectively. UF experiments were performed under the operating pressure of 120 KPa, and the change of membrane flux over time was recorded, and the CRE, TRE, PRE and SRE at different pH were analyzed with the LRE.

3.3.1. The Change of Membrane Flux

From Figure 4a, when the pH of pre-microfiltered broth was fixed, the membrane flux gradually decreased over time. In the first 50 min, the average membrane fluxes at pH 5–8 were 12.3 L/(m^2·h), 12.8 L/(m^2·h), 13. 49 L/(m^2·h), 13.77 L/(m^2·h), respectively. Membrane flux increased with the increase in pH. This may have been partly because with the increase in pH, the repulsion between the functional groups of the membrane increased, resulting in a certain expansion of the membrane pores, resulting in a larger membrane flux. It was reported that with the increase in the pH, the membranes show anionic characteristics [37]. Thus, due to membranes and impurities mostly carrying a negative charge, electrostatic repulsion is helpful for decreasing fouling [38,39]. After 50 min, it was found that the membrane flux at pH 7 and pH 8 condition began to gradually decrease, which was lower than that under other pH conditions. The reason may have been because when pH increased, some of the substances in the pre-microfiltered broth, such as proteins, formed aggregates due to attractions of calcium complexes with phosphates and/or lactate and were deposited on the membrane surface, thereby reducing the membrane flux [40].

However, in general, although the membrane flux at different pH conditions varied, the gap was not obvious.

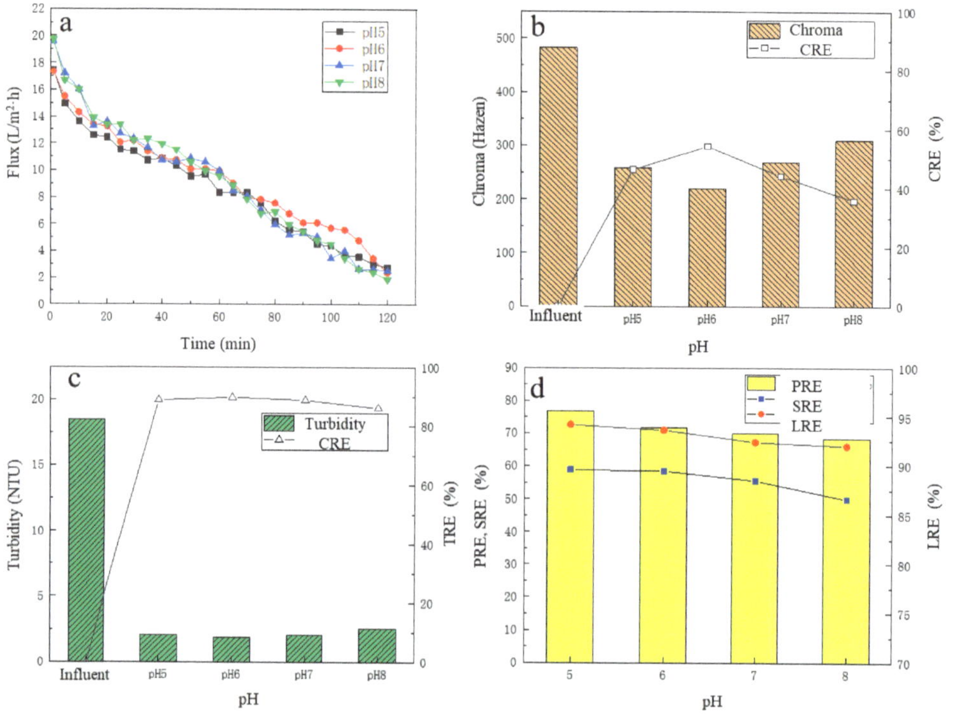

Figure 4. The effect of different pH on filtration performance: (**a**) the flux change; (**b**) the chroma removal; (**c**) the turbidity removal; (**d**) the protein removal efficiency, the total sugar removal efficiency and the lactic acid recovery efficiency.

3.3.2. The Removal of Chroma and Turbidity

Before and after filtration, the chroma changes of the pre-microfiltered broth are shown in Figure 4b. With the increase in pH, the chroma of the filtrate showed an upward trend, and the overall CRE showed a downward trend. At pH 6.0, the chroma of the filtrate was the lowest, i.e., only 220.2 Hazen, and the CRE reached 54.33%. Under different pH conditions, the turbidity changes are shown in Figure 4c. The turbidity of the pre-microfiltered broth was about 20 NTU, and the turbidity of the filtrate was basically 2.5 NTU or less. With the increase in pH of the pre-microfiltered broth, the turbidity of the filtrate showed an overall upward trend, and the TRE showed a downward trend. At pH 6.0, the turbidity of the filtrate was the lowest, only 1.89 NTU, and the TRE was 89.76%.

3.3.3. The Removal of Protein, Total Sugar and the LA Recovery

Under different pH conditions, the refinement performance of UF membrane for the pre-microfiltered broth is shown in Figure 4d. With the increase in pH, PRE and SRE showed an overall downward trend, but the decline was relatively small. At pH 5, the PRE and SRE were the highest with the values of 76.84% and 58.86%, respectively, which indicated that acidic conditions facilitated the removal of protein and total sugar. With the increase in the pH, the LRE showed a downward trend. The LRE was 94.20% at pH 5.0 and was 92.00% at pH 8.0. This may be because as the pH increased, most of the LA was presented with the form of ions, which were easily repelled by negatively charged PES membranes, thereby increasing the LA loss [41]. By comprehensively considering TRE,

CRE, PRE, SRE and LRE, it was found that lower pH conditions were more favorable for the UF performance. While considering that under low pH condition, the UF membrane flux would also decrease to a certain extent, and for achieving low the pH, the cost of hydrochloric acid regulation would increase, and so, the final choice was determined at pH 6.0 as the optimal pH condition for UF treatment.

3.4. The Effect of Cleaning Method on Polluted Membrane Recovery

3.4.1. The Influence of Cleaning Mode on Cleaning Performance

In this study, deionized water was first used to clean the contaminated membrane, and the cleaning performance of four cleaning methods on the contaminated membrane was investigated. The four cleaning methods were as follows: (i) forward pressurized cleaning for 3 min, (ii) reverse pressurized cleaning for 3 min, (iii) forward pressurized cleaning for 1.5 min + reverse pressurized cleaning for 1.5 min and (iv) reverse pressurized cleaning for 1.5 min + forward pressurized washing 1.5 min. The pressure used for cleaning was 100 KPa. The cleaning performance of the four cleaning methods on contaminated UF membrane are shown in Figure 5.

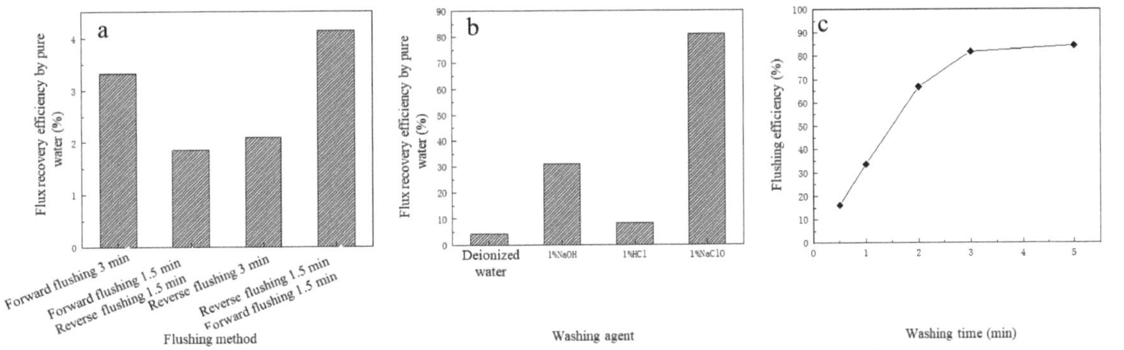

Figure 5. The effect of different flushing operations on flux recovery efficiency of microfiltration: (**a**) the flushing methods; (**b**) the flushing agents; (**c**) the effect of washing time on flushing efficiency.

From Figure 5a, it was obvious that with the same cleaning time, forward pressurized cleaning had a better effect than that of reverse pressurized cleaning to relieve membrane pollution, and the cleaning effect of forward pressurized cleaning + reverse pressurized cleaning was the worst, slightly lower than that of reverse pressurized cleaning. The reverse pressurized cleaning + forward pressurized cleaning had the best cleaning performance with the FRE of 4.13%. However, in general, regardless of which kind of cleaning method was adopted, the relief of membrane pollution was not desirable, and the maximum FRE was only 4.13%. Therefore, additional cleaning agents should be considered for enhanced cleaning effect.

3.4.2. The Influence of Cleaning Agent on Cleaning Performance

Four cleaning agents were selected to further clean the contaminated membrane. It was reported that the concentration of cleaning chemicals is important to maintain the optimum reaction rate and overcome the mass transfer resistance of fouling layers [42]. The four cleaning agents were deionized water, 1% HCl, 1% NaOH and 1% NaClO; the cleaning method was first deionized water forward pressure cleaning for 3 min, then cleaning agent reverse pressurized for 3 min with the cleaning pressure of 100 KPa. The performance of four kinds of cleaning agents on UF membrane pollution are shown in Figure 5b. The FRE with deionized water was the worst, only 4.19%, which was 50% of that with HCl. The cleaning performance of NaOH was obviously better than that with HCl, and the FRE was 31.17%. NaClO had the best cleaning effect with FRE reaching 81.19%, which was much higher than the other three cleaning agents. The reason may be that as membrane pollution was mainly caused by organic matter, the NaClO pair had strong oxidation, which could oxidize and remove most organic pollutants [43]. Therefore, 1% NaClO was finally selected as the cleaning agent. In addition, it should be mentioned that while chemical cleaning is an effective operation, it may be harsh to the membrane and damage the membrane material, resulting in reduced membrane lifetime and selectivity [44].

3.4.3. The Influence of Cleaning Time on Cleaning Performance

After the cleaning agent was selected with 1% NaClO, to further improve the cleaning efficiency, this section investigated the effect of cleaning time on cleaning effect. As shown in Figure 5c, within the first 3 min of cleaning, the FRE increased significantly with the increase in cleaning time, and when the cleaning time was 30 s, the FRE was only 16.22%, and when the cleaning time reached 3 min, the FRE increased to 81.19%. However, when the cleaning time further increased to 5 min, the FRE was 84.43%, which was slightly higher than that at 3 min. It was also reported that FRE increased by increasing the cleaning time. However, the FRE slowly decreased due to the limited chemicals to dissolve the deposited layer [45]. The other study also showed that FRE did not depend on the cleaning time, and short intensive cleaning was better than long comprehensive cleaning [46]. Therefore, to save cleaning time and the amount of cleaning agent, the more reasonable cleaning time in NaClO cleaning method was determined to be 3 min.

3.4.4. Membrane Surface Morphology Analysis

As shown in Figure 6, the surface morphology changes of the UF membrane observed through scanning electron microscopy were determined. The surface of the new UF membrane was smooth and basically free of particulate matter. There were obviously much particulate contaminants on the surface of the contaminated UF membrane, and most of the pollutants had a small particle size of about 0.5 μm. After cleaning with 1% NaClO solution, the surface pollution of the membrane was basically relieved, which fully indicated that NaClO had a significant removal effect on the surface of the UF membrane. No matter the dead-end or the crossflow UF, the membrane fouling was related to the feed feature, such as the impurity's concentration, and the filtration time. In this research, although the dead-end UF was adopted, the magnetic stirring component was equipped to relief the fouling. The increase in the concentration of solution was reflected in the membrane performance, which was same to the feed with high concentration.

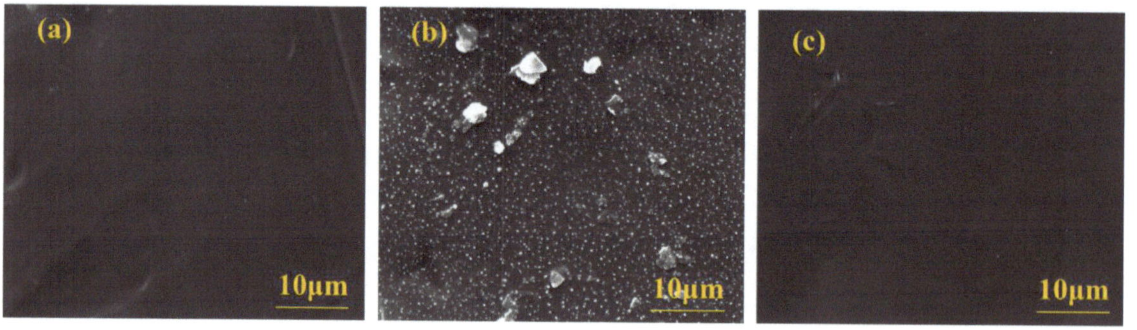

Figure 6. The surface observation of the microfiltration membrane by SEM instrument: (**a**) the new membrane; (**b**) after the blocking; (**c**) after the flushing with 1% NaClO.

3.5. *The Effect of Cleaning Cycle on UF Membrane Performance*

After the cleaning agent and cleaning method were determined, the 10 cycles of UF experiment were carried out. Each cycle ran for 30 min and the change of membrane flux over time was recorded. After the filtration was completed, the UF membrane was rinsed with 1% NaClO. The pure water fluxes of the UF membrane in each cycle were recorded, the treatment performance of each filtration cycle was analyzed, as shown in Figure 7a–e.

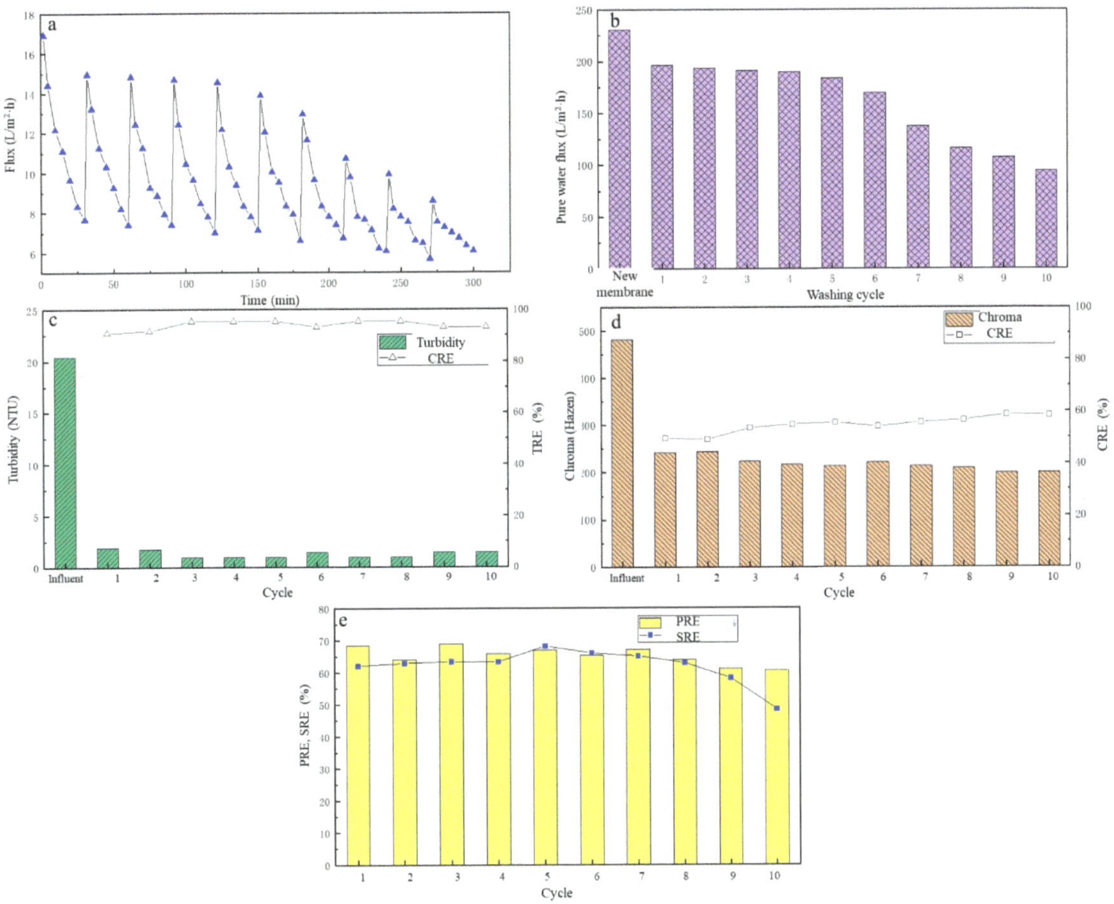

Figure 7. The effect of different flushing cycles on filtration performance: (**a**) the flux change; (**b**) the pure water flux; (**c**) the turbidity removal; (**d**) the chroma removal; (**e**) the protein removal efficiency and the total sugar removal efficiency.

From Figure 7a, the membrane flux decreased with the increase in the filtration cycle. However, from circle two to six, the membrane flux decreased significantly. The average membrane flux was maintained at around 10.1 L/(m^2·h), the effectiveness of the cleaning effect of NaClO was fully demonstrated. However, from the seventh cycle, the membrane flux began to decline sharply, and the trend of the pure water flux of the membrane was also the same, as shown in Figure 7b. When the cycles of membrane cleanings were less than six, the pure water flux was basically maintained at 180–190 L/(m^2·h). However, when the cycles of cleanings were more than six, the pure water flux of the membrane began to decline rapidly. After 10 cycles, the membrane had a pure water flux of only 94 L/(m^2·h), which was 41% of the pure water flux of the new membrane. The reason for this may be that NaClO caused some damage to the membrane pores, resulting in a decrease in membrane flux [44].

Under different filtration cycles, the TRE and CRE are shown in Figure 7c,d. After filtration, the turbidity of the filtrate was basically below 2 NTU. Although the TRE fluctuated, it was basically between 90% and 95%, and the variation was small. The CRE showed an overall upward trend with the increase of the cycles, basically maintaining between 50 and 60%, and in the ninth cycle, the CRE reached the highest value of 58.61%.

Under different filtration cycles, the PREs and SREs from the fermentation broth are shown in Figure 7e. The PREs were basically between 64% and 69% in the first eight cycles. From the eighth cycle, the PRE began to gradually decrease. The SRE increased first and then decreased with the increase in the cycle, reaching the highest in the fifth cycle, which was 68.24%. Overall, the change was small in the first eight cycles with the value of about 64%. In the 10th cycle, it had the lowest PRE and SRE of 60.57% and 48.4%, respectively. After comprehensive analysis, after six cycles, the membrane could still maintain a good membrane flux and treatment performance.

3.6. The Whole Evaluation of the UF

The whole removal performance of the UF process in this study was summarized, as shown in Figure 8. For all the impurities, UF process achieved the best removal performance with TRE of above 89.8%, followed by the CRE of 54.3%, PRE of above 71.7%, and SRE of 58.5%. The final UF filtrate had the LA content of 55.0 g/L, the chroma of 203.7 Hazen, the turbidity of 1.0 NTU, the protein content of 0.02 g/L and the total sugar content of 0.9 g/L. From these, it was obvious that for the refinement of the LA from the pre-microfiltered broth, the UF process could effectively achieve the target to some extent. The chroma was still high, which indicated that the combination of UF with other post-treatment processes is needed. In addition, the LA loss was about 6.4%. Thus, in future research, the effort to further lessen the loss of LA during the filtration should also be mentioned.

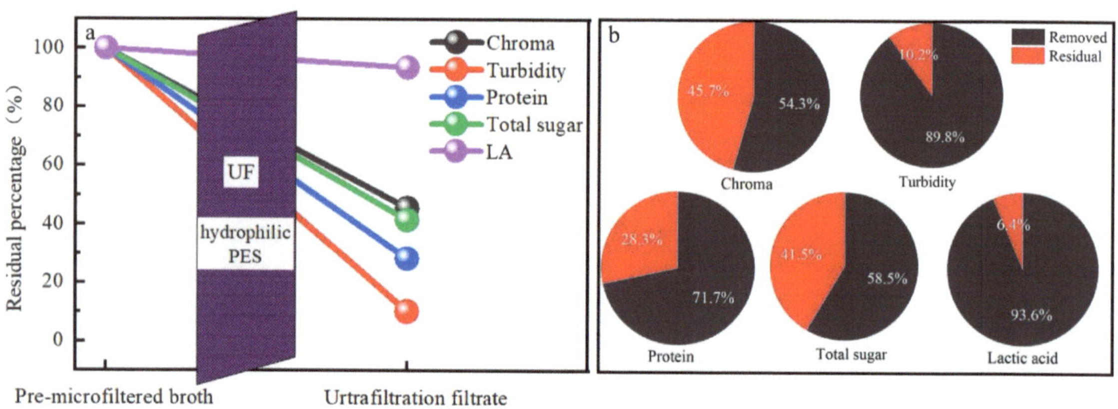

Figure 8. The whole removal performance of the microfiltration membrane separation (**a**) and the removal percentage of each component (**b**).

4. Conclusions

In this study, for the LA refinement from pre-microfiltered fermentation broth, with the 50 KDa polyethersulfone membrane, the optimum operation parameters were determined as the pressure of 120 KPa, the pH of 6.0 and the flushing mode of backwashing with 1% NaClO for 3 min. The best performance was achieved with the CRE, TRE, PRE and SRE of 54.3%, 89.8%, 71.7% and 58.5%, respectively, and LRE was 93.6%. This study paves the way to LA refinement with UF membrane technology. The obtained results had the meaning as the reference for the future research. In conclusion, the UF process is promising as the effective separation method for the preliminary refinement of fermentation broth through combing with other pre-treatment processes and post-treatment processes.

Author Contributions: Data curation, C.L. and H.Z.; formal analysis, C.L. and H.Z.; funding acquisition, Q.W. and X.S.; investigation, C.L. and H.Z.; methodology, Y.G. and X.W.; project administration, X.W. and M.G.; resources, C.L. and H.Z.; software, C.L. and X.W.; writing—original draft, Y.G. and X.S.; writing—review and editing, Y.G. and X.S. All authors have read and agreed to the published version of the manuscript.

Funding: This research was supported by the National Natural Science Foundation of China (Grant NO. 51978047 & 5217100753) and the National Key R&D Program of China (Grant NO. 2019YFC1906302 & 2019YFC1906304).

Institutional Review Board Statement: Not applicable.

Data Availability Statement: A statement was submitted.

Conflicts of Interest: The authors declare no conflict of interest.

References

1. Huang, Y.L.; Wu, Z.; Zhang, L.; Ming Cheung, C.; Yang, S.-T. Production of Carboxylic Acids from Hydrolyzed Corn Meal by Immobilized Cell Fermentation in a Fibrous-Bed Bioreactor. *Bioresour. Technol.* **2002**, *82*, 51–59. [CrossRef] [PubMed]
2. Lee, Y.; Lee, J.H.; Yang, H.J.; Jang, M.; Kim, J.R.; Byun, E.-H.; Lee, J.; Na, J.-G.; Kim, S.W.; Park, C. Efficient Simultaneous Production of Biodiesel and Glycerol Carbonate via Statistical Optimization. *J. Ind. Eng. Chem.* **2017**, *51*, 49–53. [CrossRef]
3. Berhanu, M.; Jabasingh, S.A.; Kifile, Z. Expanding Sustenance in Ethiopia Based on Renewable Energy Resources—A Comprehensive Review. *Renew. Sustain. Energy Rev.* **2017**, *75*, 1035–1045. [CrossRef]
4. Ma, X.; Gao, M.; Liu, S.; Li, Y.; Sun, X.; Wang, Q. An Innovative Approach for Reducing the Water and Alkali Consumption in the Lactic Acid Fermentation via the Reuse of Pretreated Liquid. *Bioresour. Technol.* **2022**, *352*, 127108. [CrossRef] [PubMed]
5. Xu, M.; Sun, H.; Yang, M.; Xie, D.; Sun, X.; Meng, J.; Wang, Q.; Wu, C. Biodrying of Biogas Residue through a Thermophilic Bacterial Agent Inoculation: Insights into Dewatering Contribution and Microbial Mechanism. *Bioresour. Technol.* **2022**, *355*, 127256. [CrossRef] [PubMed]
6. Zacharof, M.-P.; Lovitt, R.W. Recovery of Volatile Fatty Acids (VFA) from Complex Waste Effluents Using Membranes. *Water Sci. Technol.* **2013**, *69*, 495–503. [CrossRef]
7. Aghapour Aktij, S.; Zirehpour, A.; Mollahosseini, A.; Taherzadeh, M.J.; Tiraferri, A.; Rahimpour, A. Feasibility of Membrane Processes for the Recovery and Purification of Bio-Based Volatile Fatty Acids: A Comprehensive Review. *J. Ind. Eng. Chem.* **2020**, *81*, 24–40. [CrossRef]
8. Raposo, F.; Borja, R.; Cacho, J.A.; Mumme, J.; Orupõld, K.; Esteves, S.; Noguerol-Arias, J.; Picard, S.; Nielfa, A.; Scherer, P.; et al. First International Comparative Study of Volatile Fatty Acids in Aqueous Samples by Chromatographic Techniques: Evaluating Sources of Error. *TrAC Trends Anal. Chem.* **2013**, *51*, 127–143. [CrossRef]
9. Fernández, R.; Dinsdale, R.M.; Guwy, A.J.; Premier, G.C. Critical Analysis of Methods for the Measurement of Volatile Fatty Acids. *Crit. Rev. Environ. Sci. Technol.* **2016**, *46*, 209–234. [CrossRef]
10. Masse, L.; Massé, D.I.; Pellerin, Y. The Effect of PH on the Separation of Manure Nutrients with Reverse Osmosis Membranes. *J. Memb. Sci.* **2008**, *325*, 914–919. [CrossRef]
11. Huang, C.; Xu, T.; Zhang, Y.; Xue, Y.; Chen, G. Application of Electrodialysis to the Production of Organic Acids: State-of-the-Art and Recent Developments. *J. Memb. Sci.* **2007**, *288*, 1–12. [CrossRef]
12. de-Bashan, L.E.; Bashan, Y. Recent Advances in Removing Phosphorus from Wastewater and Its Future Use as Fertilizer (1997–2003). *Water Res.* **2004**, *38*, 4222–4246. [CrossRef] [PubMed]
13. Gluszcz, P.; Jamroz, T.; Sencio, B.; Ledakowicz, S. Equilibrium and Dynamic Investigations of Organic Acids Adsorption onto Ion-Exchange Resins. *Bioprocess. Biosyst. Eng.* **2004**, *26*, 185–190. [CrossRef]
14. Senol, A.; Dramur, U. Predicting Liquid–Liquid Equilibria of Amine Extraction of Carboxylic Acid Through Solvation Energy Relation. *Solvent Extr. Ion Exch.* **2004**, *22*, 865–883. [CrossRef]
15. Mumtaz, T.; Abd-Aziz, S.; Aini, N.; Rahman, A.; Yee, P.; Shirai, Y.; Hassan, M. Pilot-Scale Recovery of Low Molecular Weight Organic Acids from Anaerobically Treated Palm Oil Mill Effluent (POME) with Energy Integrated System. *Afr. J. Biotechnol.* **2008**, *7*, 3900–3905.
16. Lo, K.-M.; Chien, I.-L. Efficient Separation Method for Tert-Butanol Dehydration via Extractive Distillation. *J. Taiwan Inst. Chem. Eng.* **2017**, *73*, 27–36. [CrossRef]
17. Joglekar, H.G.; Rahman, I.; Babu, S.; Kulkarni, B.D.; Joshi, A. Comparative Assessment of Downstream Processing Options for Lactic. *Acid. Sep. Purif. Technol.* **2006**, *52*, 1–17. [CrossRef]
18. Manzak, A.; Kurşun, C.; Yıldız, Y. Characterization of Humic Acid Extracted from Aqueous Solutions with Polymer Inclusion Membranes. *J. Taiwan Inst. Chem. Eng.* **2017**, *81*, 14–20. [CrossRef]
19. Trad, Z.; Akimbomi, J.; Vial, C.; Larroche, C.; Taherzadeh, M.J.; Fontaine, J.-P. Development of a Submerged Anaerobic Membrane Bioreactor for Concurrent Extraction of Volatile Fatty Acids and Biohydrogen Production. *Bioresour. Technol.* **2015**, *196*, 290–300. [CrossRef]

20. Bastrzyk, J.; Gryta, M.; Karakulski, K. Fouling of Nanofiltration Membranes Used for Separation of Fermented Glycerol Solutions. *Chem. Pap.* **2014**, *68*, 757–765. [CrossRef]
21. Bonnélye, V.; Guey, L.; del Castillo, J. UF/MF as RO Pre-Treatment: The Real Benefit. *Desalination* **2008**, *222*, 59–65. [CrossRef]
22. Karakulski, K.; Gryta, M.; Bastrzyk, J. Treatment of Effluents from a Membrane Bioreactor by Nanofiltration Using Tubular Membranes. *Chem. Pap.* **2013**, *67*, 1164–1171. [CrossRef]
23. Tomczak, W.; Gryta, M. The Application of Ultrafiltration for Separation of Glycerol Solution Fermented by Bacteria. *Pol. J. Chem. Technol.* **2013**, *15*, 115–120. [CrossRef]
24. Mao, C.; Feng, Y.; Wang, X.; Ren, G. Review on Research Achievements of Biogas from Anaerobic Digestion. *Renew. Sustain. Energy Rev.* **2015**, *45*, 540–555. [CrossRef]
25. Wang, Q.; Li, H.; Feng, K.; Liu, J. Oriented Fermentation of Food Waste towards High-Value Products: A Review. *Energies* **2020**, *13*, 5638. [CrossRef]
26. Dahiya, S.; Sarkar, O.; Swamy, Y.v.; Venkata Mohan, S. Acidogenic Fermentation of Food Waste for Volatile Fatty Acid Production with Co-Generation of Biohydrogen. *Bioresour. Technol.* **2015**, *182*, 103–113. [CrossRef]
27. Gryta, M.; Tomczak, W. Microfiltration of Post-Fermentation Broth with Backflushing Membrane Cleaning. *Chem. Pap.* **2015**, *69*, 544–552. [CrossRef]
28. Sun, X.; Wang, Q.; Zhao, W.; Ma, H.; Sakata, K. Extraction and Purification of Lactic Acid from Fermentation Broth by Esterification and Hydrolysis Method. *Sep. Purif. Technol.* **2006**, *49*, 43–48. [CrossRef]
29. Gao, Z.; Ma, Y.; Liu, Y.; Wang, Q. Waste Cooking Oil Used as Carbon Source for Microbial Lipid Production: Promoter or Inhibitor. *Environ. Res.* **2022**, *203*, 111881. [CrossRef]
30. Puro, L.; Kallioinen, M.; Mänttäri, M.; Natarajan, G.; Cameron, D.; Nyström, M. Performance of RC and PES Ultrafiltration Membranes in Filtration of Pulp Mill Process Waters. *Desalination* **2010**, *264*, 249–255. [CrossRef]
31. Charcosset, C. Classical and Recent Applications of Membrane Processes in the Food Industry. *Food Eng. Rev.* **2021**, *13*, 322–343. [CrossRef]
32. Castro-Muñoz, R.; Díaz-Montes, E.; Cassano, A.; Gontarek, E. Membrane Separation Processes for the Extraction and Purification of Steviol Glycosides: An Overview. *Crit. Rev. Food Sci. Nutr.* **2021**, *61*, 2152–2174. [CrossRef] [PubMed]
33. Adikane, H.; Thakar, D.; Nene, S. Optimisation of Colour and Sugar Rejection of Black Liquor Using Membranes. *Sep. Purif. Technol.* **2004**, *36*, 229–234. [CrossRef]
34. Sadr, S.M.K.; Saroj, D.P. Membrane Technologies for Municipal Wastewater Treatment. In *Advances in Membrane Technologies for Water Treatment*; Basile, A., Cassano, A., Rastogi, N.K., Eds.; Woodhead Publishing Series in Energy; Woodhead Publishing: Oxford, UK, 2015; pp. 443–463. ISBN 978-1-78242-121-4.
35. Li, S.; Chen, H.; Zhao, X.; Lucia, L.A.; Liang, C.; Liu, Y. Impact Factors for Flux Decline in Ultrafiltration of Lignocellulosic Hydrolysis Liquor. *Sep. Purif Technol.* **2020**, *240*, 116597. [CrossRef]
36. Hidalgo, A.M.; Gómez, M.; Murcia, M.D.; Serrano, M.; Rodríguez-Schmidt, R.; Escudero, P.A. Behaviour of Polysulfone Ultrafiltration Membrane for Dyes Removal. *Water Sci. Technol.* **2018**, *77*, 2093–2100. [CrossRef]
37. Díaz-Montes, E.; Castro-Muñoz, R. Metabolites Recovery from Fermentation Broths via Pressure-Driven Membrane Processes. *Asia-Pac. J. Chem. Eng.* **2019**, *14*, e2332. [CrossRef]
38. Huang, S.; Ras, R.H.A.; Tian, X. Antifouling Membranes for Oily Wastewater Treatment: Interplay between Wetting and Membrane Fouling. *Curr. Opin. Colloid Interface Sci.* **2018**, *36*, 90–109. [CrossRef]
39. Madaeni, S.S.; Sharifnia, S.; Moradi, G. Chemical Cleaning of Microfiltration Membranes Fouled by Whey. *J. Chin. Chem. Soc.* **2001**, *48*, 179–191. [CrossRef]
40. Chandrapala, J.; Duke, M.C.; Gray, S.R.; Weeks, M.; Palmer, M.; Vasiljevic, T. Strategies for Maximizing Removal of Lactic Acid from Acid Whey—Addressing the Un-Processability Issue. *Sep. Purif. Technol.* **2017**, *172*, 489–497. [CrossRef]
41. Koschuh, W.; Thang, V.H.; Krasteva, S.; Novalin, S.; Kulbe, K.D. Flux and Retention Behaviour of Nanofiltration and Fine Ultrafiltration Membranes in Filtrating Juice from a Green Biorefinery: A Membrane Screening. *J. Memb. Sci.* **2005**, *261*, 121–128. [CrossRef]
42. Paugam, L.; Delaunay, D.; Diagne, N.W.; Rabiller-Baudry, M. Cleaning of Skim Milk PES Ultrafiltration Membrane: On the Real Effect of Nitric Acid Step. *J. Memb. Sci.* **2013**, *428*, 275–280. [CrossRef]
43. Hou, L.; Gao, K.; Li, P.; Zhang, X.; Wang, Z.; Song, P.; Yao, W. A Kinetic Model for Calculating Total Membrane Fouling Resistance in Chemical Cleaning Process. *Chem. Eng. Res. Des.* **2017**, *128*, 59–72. [CrossRef]
44. Bird, M.R.; Bartlett, M. Measuring and Modelling Flux Recovery during the Chemical Cleaning of MF Membranes for the Processing of Whey Protein Concentrate. *J. Food Eng.* **2002**, *53*, 143–152. [CrossRef]
45. Madaeni, S.; Tavakolian, H.; Rahimpour, F. Cleaning Optimization of Microfiltration Membrane Employed for Milk Sterilization. *Sep. Sci. Technol.* **2011**, *46*, 571–580. [CrossRef]
46. Naim, R.; Levitsky, I.; Gitis, V. Surfactant Cleaning of UF Membranes Fouled by Proteins. *Sep. Purif. Technol.* **2012**, *94*, 39–43. [CrossRef]

membranes

Article

The Performance of Microfiltration Process for Purifying Lactic Acid in the Fermented Broth of Kitchen Waste

Yan Guo [1], Chenglong Li [1], Hongjun Zhao [1], Ming Gao [1,*] and Qunhui Wang [1,2,*]

[1] Department of Environmental Engineering, University of Science and Technology Beijing, 30 Xueyuan Road, Haidian District, Beijing 100083, China
[2] Department of Environmental Engineering, Tianjin College, University of Science and Technology Beijing, Tianjin 301830, China
* Correspondence: gaoming402@ustb.edu.cn (M.G.); wangqh59@sina.com (Q.W.)

Abstract: Fermentation broth is plentiful with lactic acid, an important chemical applied in many fields, such as food processing, the chemical industry, and cosmetics. However, the purification of the lactic acid from the broth is still troublesome, when considering the economy. This study first investigated the purification performance of microfiltration (MF) membrane technology for a fermentation broth from kitchen waste. The effect of operation pressure, broth pH, and membrane flushing mode on the membrane filtration performance were investigated. In addition, the change in filtration performance over the increase in cycle time was also investigated. The results showed that under the optimum pressure of 100 KPa, pH of 6.0, and a backflushing mode with deionized water for 3 min, the best performance was achieved, with chroma removal, turbidity removal, protein removal and total sugar removal efficiencies of 60, 92.8, 57.64 and 32.93%, respectively. The results indicated that the MF process could be a desirable broth purification process to some extent, and it is promising in actual application. The MF process combined with other post-purification processes will form the ideal process system, which should be explored in future research.

Keywords: kitchen waste; fermentation broth; lactic acid; microfiltration; membrane; flushing

check for updates

Citation: Guo, Y.; Li, C.; Zhao, H.; Gao, M.; Wang, Q. The Performance of Microfiltration Process for Purifying Lactic Acid in the Fermented Broth of Kitchen Waste. *Membranes* **2023**, *13*, 280. https://doi.org/10.3390/membranes13030280

Academic Editor: Karin Schroën

Received: 3 January 2023
Revised: 23 January 2023
Accepted: 28 January 2023
Published: 27 February 2023

1. Introduction

Lactic acid (LA) is an important chemical, which exists widely in nature and is considered the simplest low hydroxyl carboxylic acid and bulk chemical [1]. The application range of LA is very wide in areas such as food processing, the pharmaceutical industry, chemical industry, cosmetics and other industries [2]. Even in 2010, the U.S. Department of Energy issued a report that listed LA as a potential building block for the future [3]. There are two main production methods for LA, chemical synthesis, and biological fermentation [1,4,5]. Compared with chemical synthesis, biological fermentation can use low-cost biomass waste, such as kitchen waste [6], straw [7], and sludge [8] as substrates for microbial LA fermentation, which is more environmentally friendly, has low energy consumption, and can produce LA with higher optical purity [9,10]. Today, more than 90% of the world's LA production comes from biological fermentation [11].

However, the production of LA by the biological fermentation method still faces many problems and challenges. In addition to the LA in the fermentation broth, there are also a large number of bacteria, proteins, residual sugars, inorganic salts and other impurities, and after the completion of fermentation, it is necessary to use reasonable extraction technology to separate the LA [12]. The extraction and purification of LA directly affects the quality and yield of the product. In addition, the cost of LA recovery and purification accounts for 50–80% of the total cost of LA production [13,14]. The purity required for food-grade is approximately 80–90% and more than 90% for the pharmaceutical grade, which considerably affects its price [15]. Therefore, the recovery of LA from fermentation broth is considered one of the biggest challenges in LA production and a bottleneck restricting LA production. As the composition of the LA fermentation broth is complex, the traditional LA separation process is subsequently complex, in which the acid and alkali consumption is large, the labor intensity is large, and many by-products are produced, which is not friendly to the environment [13]. Four different separation and purification methods, precipitation, solvent extraction, adsorption, and membrane separation, have been extensively investigated for LA recovery from fermentation broth [1]. As the most common and conventional method, the calcium LA crystallization-acid solution process still faces the problems of low lactate recovery efficiency (LRE) [16]. While the high cost and toxicity of the solvents towards microorganisms limit their use in in situ extraction processes [17,18]. Membrane separation technology has the advantages of high efficiency, no phase change, low energy consumption [19,20]. In addition, membranes can permit synchronous fermentation and purification in a single reactor. Membranes include microfiltration (MF), ultrafiltration, nanofiltration, reverse osmosis, and electrodialysis membranes. There is the problem that impurities in broth can quickly foul all membranes. However, the extent of fouling may be far less during microfiltration [21]. To meet the needs of LA industrial production, it is imperative to develop an efficient, low-pollution, low-energy lactic acid separation technology.

Another factor affecting the final cost of LA generated through fermentation technology is the price of raw materials, which can be as high as 40% of the LA cost [22,23]. Recently, the raised price of commercialized raw material has increased the cost of LA production, and affects the final product price [1]. Consequently, industries are turning their channels to more economic alternatives, one of which, kitchen waste, has been used to produce LA [24–28]. After the fermentation of kitchen waste, the broth is centrifuged, through which most of the particle suspension and bacteria in the broth are removed. However, there are still some suspended particulate matters in the broth, and the content of total sugar, protein and other substances is still high, which requires further processing. Until now, there has been little coverage and insufficient research about the purification of LA in the fermentation broth of kitchen waste adopting membrane separation technology. As the experience and reports of other garbage fermentation broths using MF membranes to purify LA indicated the effective performance of membranes, the use of an MF membrane to purify the LA in the fermentation broth of kitchen waste was carried out in this study.

This project used an MF membrane separation process system to separate and extract the LA in the fermentation broth of kitchen waste, and mainly studied the separation performance of the MF membrane on the LA fermentation broth; the turbidity removal efficiency (TRE), color removal efficiency (CRE), total sugar removal efficiency (SRE), protein removal efficiency (PRE) and LA recovery efficiency (LRE) were investigated. Meanwhile, the effect of operating pressure, fermentation broth pH and temperature on the filtration performance were investigated for optimizing the most suitable operating conditions for the filtration process. In addition, the cleaning effect of different cleaning modes, cleaning agents and cleaning time on the MF membrane was investigated, and the most suitable MF conditions were determined. Finally, the sustainability of stable MF performance under different cleaning cycles was explored.

2. Materials and Methods

2.1. LA Fermentation Liquid for Kitchen Waste

The LA fermentation broth used in the MF processes was derived from kitchen waste fermentation by *Montessori Enterococcus* bacteria. Kitchen waste was taken from the Hongboyuan Canteen of the University of Science and Technology, Beijing, located in Beijing, China, and the collected kitchen waste was first manually removed from the hard bones, plastic bags, paper towels and other impurities, and then put into the meat grinder to stir, grind, mix well and put into the self-sealing bag, and put into the −20 °C refrigerator for freezing [29]. Before fermentation, tap water was added to the kitchen waste in a 1:1 (w/v) ratio to adjust the concentration of total solids. The LA bacteria used in the fermentation process were *Montessori Enterococcus* CGMCC 22227, the inoculation ratio was 10%, the fermentation temperature was maintained at 43 °C, and the pH was adjusted to 6.8–7.0 with NaOH solution (10 mol/L) every 12 h.

After the fermentation was completed (84 h), the raw fermentation broth was centrifuged and separated by centrifuge (Hunan Hexi Instrument Equipment Co., Ltd., Changsha, China) (speed 12,000 rpm, centrifugation 10 min). After centrifugation, the bottom sedimentary substance was removed from the raw fermentation broth, and then the upper oily substance was removed through the separating funnel to obtain the fermentation broth required for the experiment. After centrifugation, the fermentation broth had the features of a pH of 6.2–6.3, an LA content of 63.96 g/L, a color of 1137 Hazen, a turbidity of 175 NTU, a protein content of 0.21 g/L, and a total sugar content of 3.33 g/L.

2.2. Experimental Apparatus

The MF device used in this study was a filtration cup (Shandong Bona Biotechnology Co., Ltd., Jinan, China), and the device is shown in Figure 1. The filtration cup adopted a flat membrane structure, and was equipped with a magnetic stirring rod to mix the liquid in the filtration cup, and high-pressure nitrogen (Beijing Huanyu Jinghui Gas Co., Ltd., Beijing, China) was used as the driving force to pressurize the system. Polymeric membranes are the most used in MF [30]. In this study, the MF membrane was made of polyvinylidene fluoride (PVDF), with a pore diameter of 0.1 μm, a diameter of 76 mm, and an effective filtration area of 45.36 cm^2.

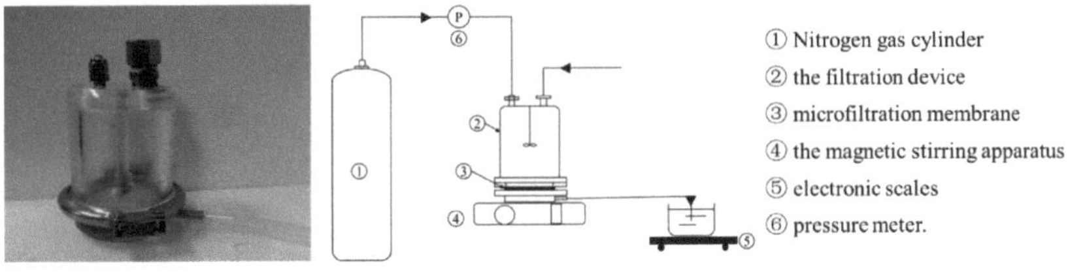

Figure 1. The illustration of the filtration configuration.

During the experiment, the MF membrane was loaded into the filtration cup, then, a certain amount of fermentation broth was loaded into the filtration cup, the magnetic stirrer below was turned on, and the speed was adjusted to 600 rpm. The nitrogen cylinder connected to the hose at the upper end of the filtration cup was opened, the pressure adjusted, and filtration performed until the outflow velocity of the liquid was less than 0.2 mL/min.

When the MF membrane was not in use, it was stored in a container filled with desiccant. After use, it was kept in a wet state in 0.5% formaldehyde solution. It was soaked in deionized water for 1 h before each use.

2.3. Analysis Indicators and Methods

2.3.1. Turbidity, Chromaticity

The turbidity and color of the fermentation broth were measured using the water quality analyzer (Beijing Lianhua Yongxing Science and Technology Development Co., Ltd., Beijing, China).

The TRE and the CRE were calculated, as shown in Equation (1) and Equation (2).

$$TRE = (C_{n0} - C_n)/C_{n0} \times 100 \quad (1)$$

where

C_n—the turbidity of the filtered transmissible fluid, NTU;

C_{n0}—turbidity of the fermentation broth before filtration, NTU.

$$CRE = (C_h - C_h)/C_{h0} \times 100 \quad (2)$$

where

C_h—the chromaticity of the filtered liquid, Hazen;
C_{h0}—the chromaticity of the fermentation broth before filtration, Hazen.

2.3.2. Determination of LA

The determination of LA in the fermentation broth was performed by high performance liquid chromatography (Shimadzu Corporation, Japan).

The calculation formula for the LRE of MF is shown in Equation (3).

$$LRE = (C_L'V')/(C_{L0}V_0) \times 100 \quad (3)$$

where

C_L'—the LA concentration of the fermented broth after filtration, g/L;
V'—product of fermented liquid after filtration, L;
C_{L0}—LA concentration of fermentation broth before filtration, g/L;
V_0—product of fermentation liquid before filtration, L.

2.3.3. Determination of Proteins

Proteins in fermentation broths were determined using the Coomassie Brilliant Blue G-250 method [31].

The PRE was calculated, as shown in Equation (4).

$$PRE = (C_{p0} - C_p)/C_{p0} \times 100 \quad (4)$$

where

C_p—the protein concentration in the filtered solution, g/L;
C_{p0}—protein concentration in fermentation broth before filtration, g/L.

2.3.4. Determination of Total Sugars

The determination of total sugars in the fermentation broth was carried out by the phenol-sulfuric acid method.

The SRE was calculated, as shown in Equation (5).

$$SRE = (C_{so} - C_s)/C_{so} \times 100 \quad (5)$$

where

C_s—total sugar concentration in the filtered permeable solution, g/L;
C_{s0}—total sugar concentration of fermentation broth before filtration, g/L.

2.3.5. Determination of Membrane FLUX

Membrane flux was defined as the volume of liquid that permeates a membrane per unit area per time. The flux calculation formula is shown in Equation (6).

$$Flux(mL/min \times m^2) = V/(t \times A) \times 100 \quad (6)$$

where

V—through the liquid product, mL;
t—filtration time, min;
A—membrane area, m^2.

2.3.6. Membrane Cleaning Effect

The membrane cleaning effect was expressed by the membrane pure water flux recovery efficiency (FRE), as shown in Equation (7).

$$FRE = (J_w - J_{fw})/(J_0 - J_{fw}) \times 100 \quad (7)$$

where

J_0— pure water permeation flux before membrane use, $L/m^2/h$;
J_{fw}—pure water permeation flux after membrane contamination, $L/m^2/h$;
Jw—pure water permeation flux after membrane cleaning, $L/m^2/h$.

3. Results and Discussion

3.1. The Effect of Operating Pressure on MF Performance

With a certain amount of fermentation broth injected into the filtration cup, the pressure was set to 60, 80, 100, 120 KPa, respectively, based on the pressure requirements (10–200 KPa) to carry out the MF separation [32]. Four sets of MF experiments were carried out to investigate the effect of different operating pressures on the removal performance of the MF membrane for turbidity, color, protein, and total sugar from fermentation broth, and LRE.

3.1.1. The Change in MF Membrane Flux

Under different operating pressures, the change in membrane flux over time is shown in Figure 2a. There was a slight trend that the higher the operating pressure, the higher the membrane flux, especially for the initial period. For the first 10 min, it could be observed that the membrane flux decreased rapidly. This is because, in the initial period, some bacteria, colloids and macromolecular particles in the fermentation liquid were quickly adsorbed to the membrane surface, then the effective membrane holes were blocked, which was manifested by a rapid decline in membrane flux [33]. For the 10–40 min, it could be observed that although the membrane flux continued to decrease over time, the downward trend decreased significantly. For 40–90 min, the downward trend of MF membrane flux was further slowed down, and the membrane flux tended to be stable, which was due to the gradual formation of an adsorption layer on the membrane surface as the filtration proceeded [34,35]. After 90 min, the membrane flux began to decrease over time, which may be due to the high concentration of pollutants in the remaining fermentation broth, which further aggravated the membrane pollution and thus decreased the membrane flux. The average membrane fluxes under the four pressures (from high to low) were 6.68, 6.68, 6.41 and 5.94 $L/(m^2 \cdot h)$, respectively. It was observed that the average membrane flux generally increased with the increase in the operating pressure (from 60 KPa to 100 KPa), but when the pressure had increased to a certain extent (from 100 KPa to 120 KPa), the average membrane flux was almost stable. The reason may be that with the high operating pressure, the filter cake layer accumulated on the membrane surface was compacted, resulting in an increase in the thickness and density of the filter layer, which increased the filtration resistance and caused the membrane flux to increase insignificantly [32,35].

3.1.2. The Removal Performance of Chroma and Turbidity

The color change in the fermentation broth is shown in Figure 2b, before and after the MF treatment. Before MF, the color of the fermentation broth was 1137 Hazen, and after the MF treatment, the color of the filtrate was significantly reduced, basically maintained between 440 and 480 Hazen, and the CRE was about 60%. With the increase in the operating pressure, the color of the filtrate showed an overall upward trend, but the change was minor. The minimum chromaticity was 441.7 Hazen at the operating pressure of 60 KPa. At the operating pressure of 120 KPa, the maximum chromaticity was 492.5 Hazen. Therefore, the MF treatment adopted in this study could effectively reduce the turbidity of the fermentation broth.

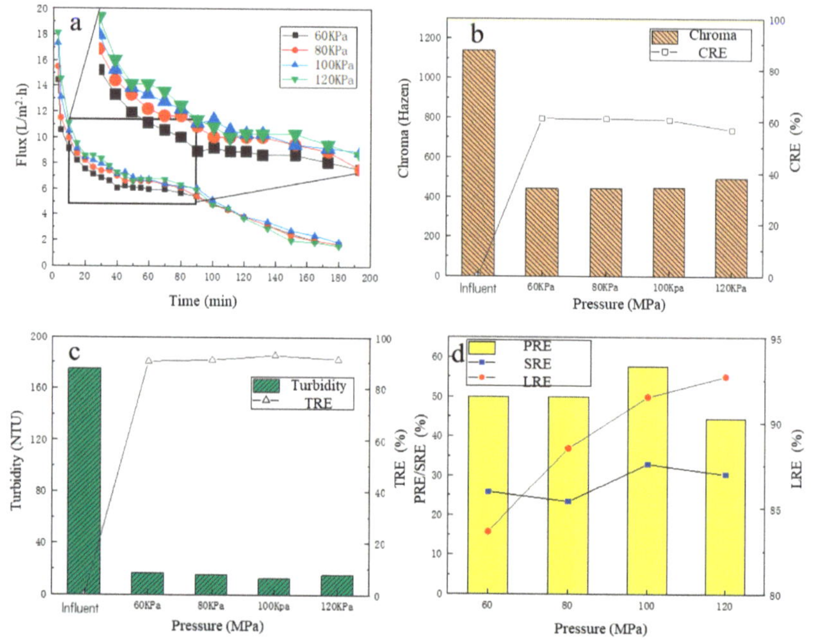

Figure 2. The effect of different pressures on filtration performance: (**a**) the flux change; (**b**) the chroma removal; (**c**) the turbidity removal; (**d**) the protein removal efficiency, the total sugar removal efficiency, and the lactic acid recovery efficiency.

After centrifugation, most of the solid suspensions in the fermentation broth were removed, but many particle suspensions remained in the fermentation broth, and the turbidity of the fermentation liquid was still high (about 175NTU). Under different operating pressures, the turbidity of the filtrate and the TRE are shown in Figure 2c. After MF treatment, the turbidity of the filtrate decreased significantly, basically below 20 NTU, and the TRE reached more than 90%. It was obvious that the different operation pressures had a weak effect on the TRE, and at the operating pressure of 100 KPa, the filtrate turbidity was the lowest, only 12.61 NTU, and the TRE was 92.79%. At the operating pressure of 60 KPa, the filtrate turbidity was the highest, which was only 16.75 NTU, and the TRE reached 90.43%, which fully demonstrated that the MF operation could effectively reduce the turbidity of the fermentation broth.

3.1.3. The Removal Performance of Protein and Total Sugar and Lactic Recovery Performance

Under four operating pressure conditions, the PRE and SRE are shown in Figure 2d. With the increase in pressure, the PRE increased. The PRE reached the highest value of 57.64% at 100 KPa, while the SRE had a fluctuating increase trend, and at 100 KPa, the SRE reached the maximum value of 32.93%. The reason why the PRE was higher than the SRE was partly because proteins, as hydrophobic compounds, are more readily adsorbed onto hydrophobic membrane surfaces than hydrophilic solutes due to hydrophobic interactions [36]. For the LRE, with the increase in pressure, the LRE increased significantly, and at 60 KPa, the LRE was only 83.63%, while at the pressure of 100 KPa, the LRE increased to 91.51%. This is because, with the increase in pressure, the share of filtrated flux also increased accordingly, resulting in clearer liquid, so its LA loss was lower. When the operating pressure continued to rise to 120 KPa, the LRE further increased to 92.70%, which was not a significant improvement, consistent with the law of change in membrane flux. Considering that the CRE, TRE, PRE and SRE of the fermentation broth were high under the operating pressure of 100 KPa with the relatively small energy consumption, the optimal operating pressure of the MF membrane was finally determined at 100 KPa.

3.2. The Effect of pH on MF Performance

In addition to the operating pressure, the pH of the fermentation broth also influences the MF performance. Four sets of MF experiments with the fermented broth placed in the filtration cup under different pHs (5, 6, 7, 8), adjusted by 6 mol/L HCl and 10 mol/L NaOH solution, and the nitrogen cylinder valve pressure of 100 KPa, were carried out. The turbidity, color, protein, total sugar, and LRE from the fermentation broth at different pHs were investigated.

3.2.1. The Change in MF Membrane Flux

Under different pH conditions, the changes in the membrane flux over time are shown in Figure 3a. Similar to Section 3.1.1, with the time proceeding, the membrane flux continued to decline, and the change trend in the membrane flux under the four pH conditions showed no obvious differences. There was a slight trend that the higher the operating pH, the higher the membrane flux, especially for the initial period. For the first 10 min, it could be observed that the membrane flux decreased rapidly. For 10–40 min, it could be observed that the membrane flux under the pH 8 condition was still the highest compared to the other three experimental sets, while the membrane flux under the pH 6 condition was slightly higher than those under the pH 5 and 7 conditions. After 40 min, the membrane flux at pH 8 decreased further and gradually decreased to less than that experienced under pHs of 5, 6, and 7. The reason may be that the pH affected the membrane–impurity interactions; it was reported that most polymers behave as cations in an acidic medium, and they have no interaction with positively charged compounds, whereas, in an alkaline medium, they have anionic behavior, having minimal interaction with negatively charged compounds [30,37,38]. Thus, in the initial stage, membranes could be used to separate compounds with the same charge based on the effect of electrostatic repulsion; then, the membrane underwent a period of slow fouling; in the later stage, the broth had higher impurity concentrations, which showed a stronger concentration gradient, which finally overcame the electrostatic repulsion, causing the rapid fouling of the membranes.

3.2.2. The Removal Performance of Chroma and Turbidity

Under different pH conditions, before and after the MF treatment, the chroma change in the fermentation broth is shown in Figure 3b. With the increase in pH of the fermentation broth, the CRE in the MF process also showed a downward trend, and under the pH of 5, the CRE was the highest at 63.3%; the chroma of the fermentation broth after MF was only 417.27 Hazen. Under different pH conditions, the turbidity of the filtrate and the TRE are shown in Figure 3c. After the MF treatment, the turbidity of the fermentation broth decreased significantly, all below 20 NTU, and the TRE reached more than 92%. With the increase in the pH of the fermentation broth, the turbidity of the filtrate showed an overall upward trend, and the TRE showed a downward trend, but the overall change was small. When the pH was 5, the TRE was highest, reaching 94.8%, and the turbidity of the filtrate was 9.07 NTU. As stated in the Section 3.2.1, under higher pH conditions, due to electrostatic repulsion, some impurities would permeate through the membrane, and then enter the filtrate, which caused the impurities in the filtrate to increase.

3.2.3. The Removal Performance of Protein and Total Sugar

Under four pH conditions, the SRE and PRE are shown in Figure 3d. With the increase in the pH of the fermentation broth, the SRE showed a gradual decreasing trend, when the pH of the fermentation broth was 5, the SRE was the highest, up to 37.99%, indicating that the MF membrane had a better removal effect on the total sugar under acidic conditions.

There was no obvious law in the change in PRE, but overall, the change was small, and the highest PRE was 60.38% with the fermentation broth of pH 6. For the LRE, with the increase in pH, the LRE had a certain decrease, and when the pH = 5, the LRE was the highest, 92.00%. When the pH increased to 8, the LRE decreased to 91.13%. Overall, however, the change in LRE was small, no more than 1%. There were some reports that the membranes based on hydrophobic materials tend to interact easily with lipids, proteins, and peptides [38,39], which could partly explain why the PRE was higher than the SRE.

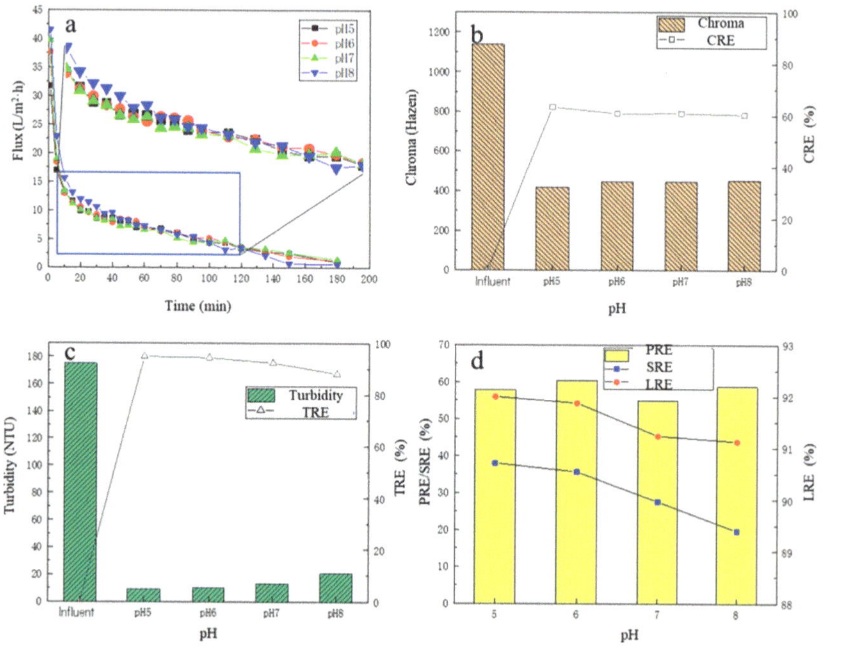

Figure 3. The effect of different pHs on filtration performance: (**a**) the flux change; (**b**) the chroma removal; (**c**) the turbidity removal; (**d**) the protein removal efficiency, the total sugar removal efficiency, and the lactic acid recovery efficiency.

Considering various indicators, the chroma, turbidity, and SRE in the fermentation broth decreased with the increase in pH, and the LRE increased slightly with the decrease in pH, so the conditions with lower pH should be selected as much as possible. However, considering that the membrane flux decreases with the decrease in pH, and the adjustment of the pH of the fermentation broth requires the addition of hydrochloric acid, for the sake of cost, it should be as close as possible to the pH of the initial fermentation broth itself (6.2). Therefore, pH 6.0 was finally selected as the best pH condition in the MF process.

3.3. The Effect of Cleaning Method on Polluted Membrane Recovery

As can be seen from Figures 2a and 3a, as the time passed by, there was a significant decrease in the membrane flux, which reflected the aggravation of the membrane blocking. In practice, to reduce the operating costs of the filtration process, the contaminated membrane should be cleaned to recover the membrane flux to some extent for prolonging the longevity of the membrane. Membrane cleaning methods can be divided into physical cleaning, chemical cleaning, biological cleaning and so on. Physical cleaning includes forward cleaning, reverse cleaning, ultrasonic cleaning, etc. Among these methods, hydraulic cleaning and gas–liquid cleaning technology are currently widely used as physical cleaning technologies.

3.3.1. The Influence of Cleaning Mode on Cleaning Performance

In this study, four kinds of cleaning modes were conducted to investigate the membrane flushing performance, and they were: 3 min forward pressurized cleaning; 3 min reverse pressurized cleaning; 1.5 min forward pressurized cleaning + 1.5 min reverse pressurized cleaning; and 1.5 min reverse pressurized cleaning + 1.5 min forward pressurized washing.

The operation process was as follows: (1) first, a brand new MF membrane was taken and soaked in deionized water for 1h, and the J_0 measured at 100 KPa; (2) the MF membrane was put into the filtration cup, 60mL of fermentation broth was added, and filtered at 100 KPa for 30 min; (3) the remaining fermentation liquid was poured out, the filtration cup was washed with deionized water, deionized water was added, the pressure adjusted to 100 KPa, and the J_{fw} measured; (4) the remaining liquid in the filtration cup was poured out, the filtration cup cleaned, deionized water added, adjusting the pressure to 100 KPa; then, it was forward pressure cleaned for 3 min and the J_w measured. The FRE was calculated according to the J_0, the J_{fw}, and the J_w based on Equation (7). The cleaning operations for the other three modes were the same as that for the forward pressure model.

The cleaning effect of the four cleaning methods on membrane MF pollution is shown in Figure 4a. The pure water FRE of forward pressurized cleaning was the worst, and the pure water FREs of forward pressurized cleaning for 1.5 min + reverse pressurized cleaning for 1.5 min mode and forward pressurized cleaning for 1.5 min + reverse pressurized cleaning for 1.5 min mode were close, but both were less effective than reverse pressurized cleaning for 3 min. The reason may be that most of the pollutants on the surface of the MF membrane were substances with large particle size, and their particle size was greater than the pore size (0.1 μm) of the MF membrane itself, with the result that in the forward pressurized cleaning mode, most of the polluting substances still stayed on the surface of the MF membrane, making the membrane pollution impossible to remove [40]. When reverse pressurized cleaning was carried out, most of the pollutants accumulated on the surface of the membrane were washed away by the water flow, and the longer the backwash time, the better the removal effect of membrane pollution. Therefore, backwashing was selected as the cleaning method of the MF membrane.

3.3.2. The Influence of Cleaning Agent on Cleaning Performance

Chemical cleaning is generally required when backwashing is unable to remove the foulants or restore the membrane flux [36]. To further improve the FRE of the MF membrane, four cleaning agents were selected to clean the contaminated MF membrane. The four cleaning agents were deionized water, 1% HCl solution, 1% NaOH solution, 1% NaClO solution; the cleaning method was reverse pressurized cleaning, the pressure was 100 KPa, and the time was 3min. The removal performance of the four kinds of cleaning agents is shown in Figure 4b.

From Figure 4b, the reverse pressurized cleaning effect of deionized water was the best, and the FRE was 66.51%, indicating that for the MF membrane used in this study, compared with chemical cleaning, the effect of physical cleaning was better. The cleaning effect of 1% HCl was second, and the FRE was 51.39%, which may have been because some insoluble substances become soluble substances under acidic conditions, so that membrane pollution was reduced. The NaClO cleaning effect was third after HCl cleaning, and the FRE was 47.51%, which may be due to NaClO solution having strong oxidation properties, so that some organic pollutants were oxidized and decomposed [41]. The NaOH cleaning effect was the worst; the reason may be that the diaphragm reacted with NaOH, affecting the membrane pore structure and flux, based on relevant studies that showed that NaOH solution will react with PVDF membrane material, causing the diaphragm to appear brownish and discolored [42]. It was reported that HCI had a moderate effect and NaOH had the weakest effect on cleaning [36,43]. On the other hand, it may be because that under the effect of NaOH, certain calcium salts and lactate salts formed complexes, which combined with organic matter such as proteins to form aggregates. Then the formed aggregates were strongly absorbed by the membrane surface or membrane pores, making the membrane flux difficult to recover [44]. Compared with HCl, NaOH and NaClO, deionized water is cheaper, and the FRE was highest, so the deionized water pressurized backwash was finally selected as the cleaning method of the MF membrane.

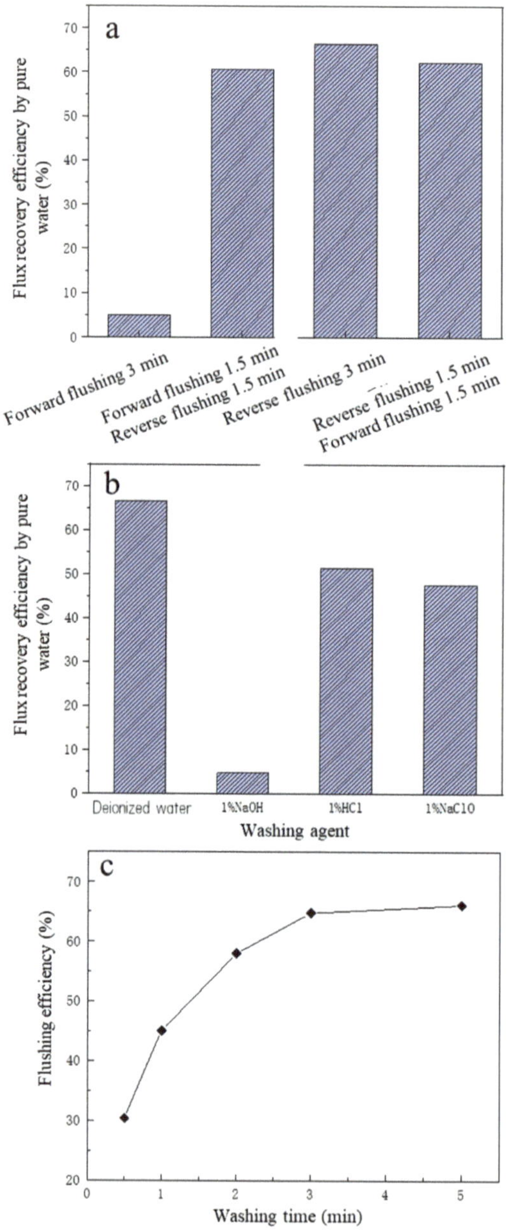

Figure 4. The effect of different flushing operations on flux recovery efficiency of microfiltration: (**a**) the flushing methods; (**b**) the flushing agents; (**c**) the effect of washing time on flushing efficiency.

3.3.3. The Influence of Cleaning Time on Cleaning Performance

In addition to the cleaning method and cleaning agent, the cleaning time also has a greater impact on the cleaning effect. After determining the cleaning agent and cleaning method (deionized water backwash), to further improve the FRE, the influence of deionized water backwash time on the cleaning effect was investigated. As can be seen from Figure 4c, within the first 3 min of deionized water backwashing, the membrane FRE increased significantly with the increase in cleaning time. When the deionized water backwashing time was 30 s, the membrane cleaning efficiency was only 30.36%, and when the cleaning time reached 3 min, the membrane cleaning efficiency rose to 66.51%, which was about doubled. However, when the cleaning time exceeded 3 min, the membrane FRE basically no longer increased with time, and the cleaning efficiency was 67.64% even when the cleaning time was 5 min. Compared with the cleaning time of 3 min, the FRE just increased by 1.3%, and the increase was weak. Therefore, to save cleaning time and cleaning water, the more reasonable cleaning time for deionized water backwashing should be 3 min.

3.3.4. MF Membrane Surface Morphology Analysis

To study the morphology of the membrane surface before and after MF membrane cleaning, the membrane surface was determined by scanning electron microscopy, and the results are shown in Figure 5. As can be seen from the figure, some loose pores could be observed on the membrane surface of the new MF. After contamination, the membrane pores became invisible due to contaminants, and some particulate matter accumulated on the membrane surface. After backwashing with deionized water, the particles on the surface of the membrane were reduced, but a small number were still present. At the same time, the membrane pores were still blocked, indicating that deionized water could only remove part of the deposits on the surface of the membrane, and the contaminants blocked in the membrane pores could not be effectively removed, so the increase in the FRE of the membrane after only washing with water was not obvious.

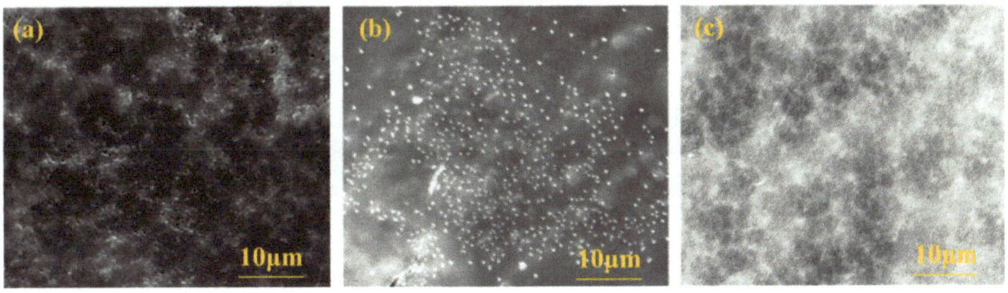

Figure 5. The surface observation of the microfiltration membrane by SEM instrument: (**a**) the new membrane; (**b**) after blocking; (**c**) after flushing with deionized water.

3.4. *The Effect of Cleaning Cycle on MF Membrane Performance*

After the MF membrane cleaning mode was determined, to further analyze the membrane flux change and treatment performance of the MF for fermentation broth after several cleaning phases, a 10-cycle MF experiment of the fermentation liquid was carried out, and each cycle was carried out for 30 min, recording the change in the membrane flux over time when the MF membrane filtered the lactate fermentation liquid in each cycle. After filtration was complete, the contaminated MF membrane was backwashed with deionized water and the pure water flux of the MF membrane was recorded after each cleaning. The treatment performance during each MF cycle was analyzed. The specific results are shown in Figure 6.

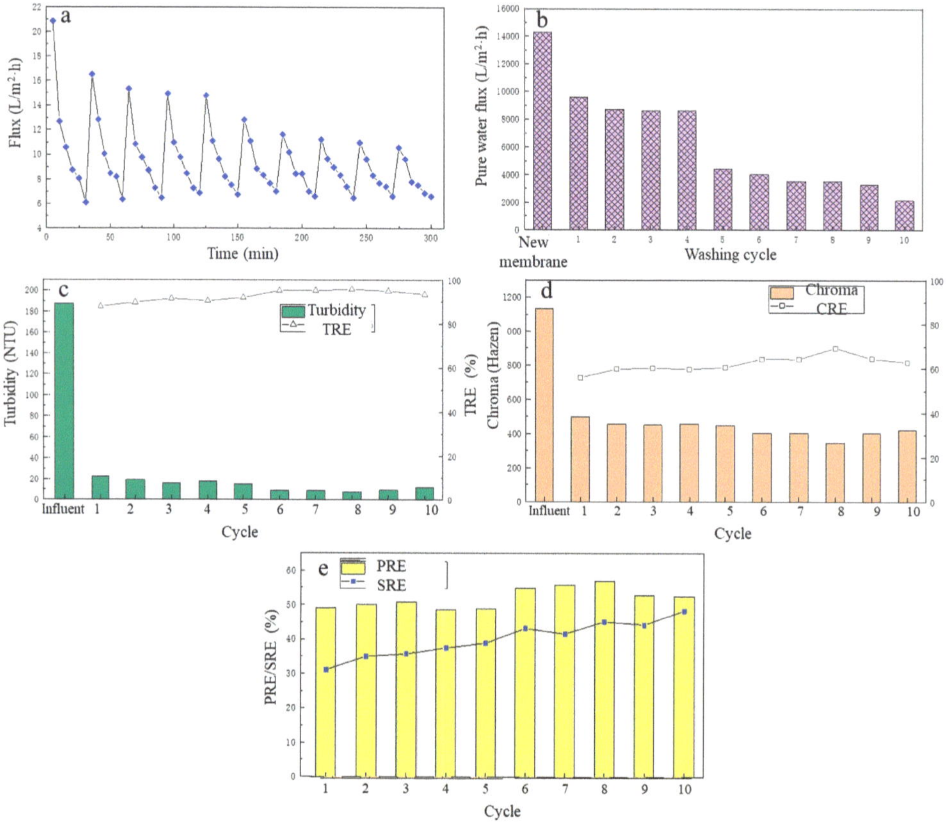

Figure 6. The effect of different flushing cycles on filtration performance: (**a**) the flux change; (**b**) the pure water flux; (**c**) the turbidity removal; (**d**) the chroma removal; (**e**) the protein removal efficiency and the total sugar removal efficiency.

From Figure 6a, after the first cleaning, the flux of the LA fermentation broth filtered by the MF membrane in the second cycle decreased to a certain extent compared with the first cycle. In the next three MF cycles (that is, from cycle 3 to cycle 5), although the MF membrane flux continued to decline, the average membrane fluxes of cycles 3–5 were 9.74 L/(m^2·h), 9.72 L/(m^2·h), 9.68 L/(m^2·h), respectively. However, in general, the MF membrane flux did not change much in the initial period, and it was basically maintained at a relatively high level, which fully explained the stability of the deionized water backwash effect. However, from the sixth cycle, the MF membrane flux began to decrease at a faster speed, and in the ninth cycle, the average membrane flux was the lowest, only 8.18 L/(m^2·h). The same was true for the change in the pure water flux of the MF membrane, as shown in Figure 6b. After the first cleaning, the pure water flux had a large decrease, and then until the fourth cleaning, the pure water flux basically did not change much. Similarly, after the fifth cleaning, the pure water flux of the MF membrane had decreased significantly, and continued to decline, and the pure water flux of the MF membrane after 10 cleaning phases was only 2169.6 L/(m^2·h). This is because the deionized water backwashing could only remove suspended solids deposited on the surface of the membrane. For the first five filtration cycles, the main cause of membrane pollution was the pollution layer accumulated on the membrane surface, and after water backwashing, the contaminated layer was effectively removed, and the membrane flux was effectively restored. However, with the increase in filtration cycles and time, the pollution inside the membrane pores gradually accumulated and was difficult to remove with the water backwash, resulting in a significant decrease in membrane flux.

Over the 10 filtration cycles, the removal of turbidity and chroma of the fermentation broth is shown in Figures 6c and 6d, respectively. In general, the TRE and CRE of the fermentation broth by the MF membrane increased with the increase in number of filtration cycles. During these cycles, the TRE was maintained between 88 and 96%, and the CRE was maintained between 56 and 70%. During the eighth filtration cycle, TRE and CRE reached their highest, at 95.78 and 69.35%, respectively.

Over 10 filtration cycles, the PRE and SRE from the fermentation broth by the MF membrane are shown in Figure 6e. In the first five filtration cycles, the PRE change was small and basically maintained at about 50%, and in the last five MF cycles, the PRE first increased with the increase in the filtration cycle, reaching a maximum 57.02 % in the eighth MF cycle. Then over the next two cycles, it gradually decreased, but still was around 52%. The SRE showed an overall upward trend with the increase in number of the filtration cycles, with the SRE of 31.11% in the first MF cycle. As the number of filtration cycles increased, it was in the 10th MF cycle that the SRE reached its highest at 48.39%. The reason for the analysis results may be that with the increase in the number of filtrations, the membrane pollution inside the MF membrane gradually accumulated, which reduced the size of the membrane pores and increased the retention rate of the membrane, thereby increasing the SRE.

3.5. The Whole Evaluation of the MF in This Study

The whole removal performance of the MF in this study was summarized, as shown in Figure 7. Among all the impurities, the MF membrane achieved the best removal performance for turbidity with a TRE of more than 94%, followed by the chroma and protein, with a removal efficiency of more than 60%. The total sugar removal performance was the worst. From these, it is obvious that for the purification of the LA in kitchen waste fermentation broth, the MF process could achieve the effect to some extent, while it was not enough to realize higher purity. The combination of MF with other post-treatment processes was needed. In addition, after the microfiltration, the LA loss was about 8.1%. Thus, in future research, the effort to lower the loss of LA during filtration should be considered.

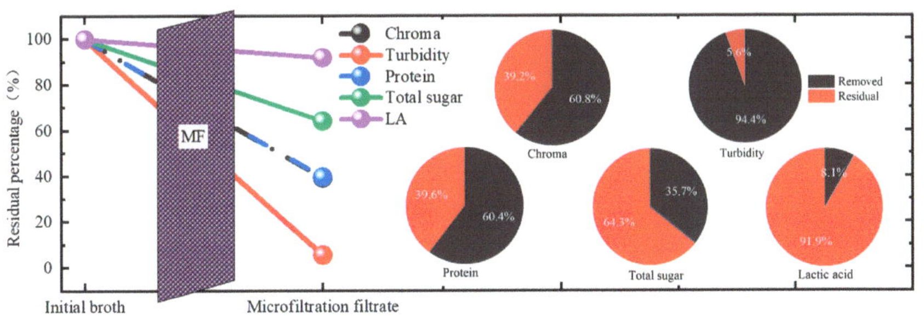

Figure 7. The whole removal performance of the microfiltration membrane separation and the removal percentage of each component.

4. Conclusions

In this study, for the purification of kitchen waste fermentation broth, the optimum operation parameters were determined as a pressure of 100 KPa, pH of 6.0, and a flushing mode of backwashing with deionized water for 3 min. The best performance was achieved with the CRE, TRE, PRE and SRE of 60, 92.8, 57.64 and 32.93%, respectively. This study paves the way for the purification of kitchen waste fermentation broth with MF membrane technology. The obtained results have meaning as a reference for future research. In all, the MF process is promising as an effective separation method for the preliminary purification of kitchen waste fermentation broth, and this study supplies the basic reference for further exploration.

Author Contributions: Data curation, C.L.; formal analysis, C.L.; funding acquisition, Q.W.; investigation, C.L. and H.Z.; methodology, Y.G.; project administration, M.G.; resources, C.L. and H.Z.; software, C.L.; writing—original draft, Y.G.; writing—review and editing, Y.G. All authors have read and agreed to the published version of the manuscript.

Funding: This research was funded by the National Key R&D Program of China (Grant NO. 2019YFC1906302 & 2019YFC1906304), the National Natural Science Foundation of China (Grant NO. 51978047 & 5217100753), and Tianjin Education Commission Scientific Research Planning Project [grant number 2021KJ064].

Institutional Review Board Statement: Not applicable.

Data Availability Statement: A statement was submitted.

Conflicts of Interest: The authors declare no conflict of interest.

References

1. Kim, J.; Kim, Y.-M.; Lebaka, V.R.; Wee, Y.-J. Lactic Acid for Green Chemical Industry: Recent Advances in and Future Prospects for Production Technology, Recovery, and Applications. *Fermentation* **2022**, *8*, 609. [CrossRef]
2. Li, C.; Gao, M.; Zhu, W.; Wang, N.; Ma, X.; Wu, C.; Wang, Q. Recent Advances in the Separation and Purification of Lactic Acid from Fermentation Broth. *Process Biochem.* **2021**, *104*, 142–151. [CrossRef]
3. Komesu, A.; Allan Rocha de Oliveira, J.; Helena da Silva Martins, L.; Regina Wolf Maciel, M.; Maciel Filho, R. Lactic Acid Production to Purification: A Review. *Bioresources* **2017**, *12*, 4364–4383. [CrossRef]
4. Cabrera-González, M.; Ahmed, A.; Maamo, K.; Salem, M.; Jordan, C.; Harasek, M. Evaluation of Nanofiltration Membranes for Pure Lactic Acid Permeability. *Membranes* **2022**, *12*, 302. [CrossRef]
5. Ma, X.; Gao, M.; Liu, S.; Li, Y.; Sun, X.; Wang, Q. An Innovative Approach for Reducing the Water and Alkali Consumption in the Lactic Acid Fermentation via the Reuse of Pretreated Liquid. *Bioresour. Technol.* **2022**, *352*, 127108. [CrossRef]
6. Ren, Y.; Yu, M.; Wu, C.; Wang, Q.; Gao, M.; Huang, Q.; Liu, Y. A Comprehensive Review on Food Waste Anaerobic Digestion: Research Updates and Tendencies. *Bioresour. Technol.* **2018**, *247*, 1069–1076. [CrossRef] [PubMed]
7. Chen, H.; Huo, W.; Wang, B.; Wang, Y.; Wen, H.; Cai, D.; Zhang, C.; Wu, Y.; Qin, P. L-Lactic Acid Production by Simultaneous Saccharification and Fermentation of Dilute Ethylediamine Pre-Treated Rice Straw. *Ind. Crops Prod.* **2019**, *141*, 111749. [CrossRef]
8. Dhandapani, B.; Vishnu, D.; Murshid, S.; Alagappan, R.P.; Muruganandh, R.; Prasanth, D.; Sekar, S.; Senthilkumar, K. Production of Lactic Acid from Industrial Waste Paper Sludge Using Rhizopus Oryzae MTCC5384 by Simultaneous Saccharification and Fermentation. *Chem. Eng. Commun.* **2021**, *208*, 822–830. [CrossRef]
9. Philp, J.C.; Ritchie, R.J.; Allan, J.E.M. Biobased Chemicals: The Convergence of Green Chemistry with Industrial Biotechnology. *Trends Biotechnol.* **2013**, *31*, 219–222. [CrossRef]
10. Abdel-Rahman, M.A.; Tashiro, Y.; Sonomoto, K. Recent Advances in Lactic Acid Production by Microbial Fermentation Processes. *Biotechnol. Adv.* **2013**, *31*, 877–902. [CrossRef]
11. Ajala, E.O.; Olonade, Y.O.; Ajala, M.A.; Akinpelu, G.S. Lactic Acid Production from Lignocellulose–A Review of Major Challenges and Selected Solutions. *ChemBioEng Rev.* **2020**, *7*, 38–49. [CrossRef]
12. Joglekar, H.G.; Rahman, I.; Babu, S.; Kulkarni, B.D.; Joshi, A. Comparative Assessment of Downstream Processing Options for Lactic Acid. *Sep. Purif. Technol.* **2006**, *52*, 1–17. [CrossRef]
13. Komesu, A.; Regina, M.; Maciel, W.; Filho, R.M. Separation and Purification Technologies for Lactic Acid—A Brief Review. *Bioresources* **2017**, *12*, 6885–6901. [CrossRef]
14. Taleghani, H.G.; Ghoreyshi, A.A.; Najafpour, G.D. Thin Film Composite Nanofiltration Membrane for Lactic Acid Production in Membrane Bioreactor. *Biochem. Eng. J.* **2018**, *132*, 152–160. [CrossRef]
15. Abdel-Rahman, M.A.; Tashiro, Y.; Sonomoto, K. Lactic Acid Production from Lignocellulose-Derived Sugars Using Lactic Acid Bacteria: Overview and Limits. *J. Biotechnol.* **2011**, *156*, 286–301. [CrossRef]
16. Pal, P.; Sikder, J.; Roy, S.; Giorno, L. Process Intensification in Lactic Acid Production: A Review of Membrane Based Processes. *Chem. Eng. Process. Process Intensif.* **2009**, *48*, 1549–1559. [CrossRef]
17. Krzyżaniak, A.; Leeman, M.; Vossebeld, F.; Visser, T.J.; Schuur, B.; de Haan, A.B. Novel Extractants for the Recovery of Fermentation Derived Lactic Acid. *Sep. Purif. Technol.* **2013**, *111*, 82–89. [CrossRef]
18. Díaz-Montes, E.; Castro-Muñoz, R. Metabolites Recovery from Fermentation Broths via Pressure-Driven Membrane Processes. *Asia-Pac. J. Chem. Eng.* **2019**, *14*, e2332. [CrossRef]
19. Połom, E.; Szaniawska, D. Rejection of Lactic Acid Solutions by Dynamically Formed Nanofiltration Membranes Using a Statistical Design Method. *Desalination* **2006**, *198*, 208–214. [CrossRef]

20. Komesu, A.; Maciel, M.R.W.; de Oliveira, J.A.R.; da Silva Martins, L.H.; Filho, R.M. Purification of Lactic Acid Produced by Fermentation: Focus on Non-Traditional Distillation Processes. *Sep. Purif. Rev.* **2017**, *46*, 241–254. [CrossRef]
21. Datta, R.; Henry, M. Lactic Acid: Recent Advances in Products, Processes and Technologies—A Review. *J. Chem. Technol. Biotechnol.* **2006**, *81*, 1119–1129. [CrossRef]
22. Cubas-Cano, E.; González-Fernández, C.; Ballesteros, M.; Tomás-Pejó, E. Biotechnological Advances in Lactic Acid Production by Lactic Acid Bacteria: Lignocellulose as Novel Substrate. *Biofuels Bioprod. Biorefin.* **2018**, *12*, 290–303. [CrossRef]
23. Reddy, L.V.; Kim, Y.-M.; Yun, J.-S.; Ryu, H.-W.; Wee, Y.-J. L-Lactic Acid Production by Combined Utilization of Agricultural Bioresources as Renewable and Economical Substrates through Batch and Repeated-Batch Fermentation of Enterococcus Faecalis RKY1. *Bioresour. Technol.* **2016**, *209*, 187–194. [CrossRef] [PubMed]
24. Kwan, T.H.; Pleissner, D.; Lau, K.Y.; Venus, J.; Pommeret, A.; Lin, C.S.K. Techno-Economic Analysis of a Food Waste Valorization Process via Microalgae Cultivation and Co-Production of Plasticizer, Lactic Acid and Animal Feed from Algal Biomass and Food Waste. *Bioresour. Technol.* **2015**, *198*, 292–299. [CrossRef]
25. Tashiro, Y.; Inokuchi, S.; Poudel, P.; Okugawa, Y.; Miyamoto, H.; Miayamoto, H.; Sakai, K. Novel PH Control Strategy for Efficient Production of Optically Active L-Lactic Acid from Kitchen Refuse Using a Mixed Culture System. *Bioresour. Technol.* **2016**, *216*, 52–59. [CrossRef]
26. Kwan, T.H.; Hu, Y.; Lin, C.S.K. Valorisation of Food Waste via Fungal Hydrolysis and Lactic Acid Fermentation with Lactobacillus Casei Shirota. *Bioresour. Technol.* **2016**, *217*, 129–136. [CrossRef]
27. Tang, J.; Wang, X.C.; Hu, Y.; Ngo, H.H.; Li, Y. Dynamic Membrane-Assisted Fermentation of Food Wastes for Enhancing Lactic Acid Production. *Bioresour. Technol.* **2017**, *234*, 40–47. [CrossRef]
28. Pleissner, D.; Neu, A.-K.; Mehlmann, K.; Schneider, R.; Puerta-Quintero, G.I.; Venus, J. Fermentative Lactic Acid Production from Coffee Pulp Hydrolysate Using Bacillus Coagulans at Laboratory and Pilot Scales. *Bioresour. Technol.* **2016**, *218*, 167–173. [CrossRef]
29. Xu, M.; Sun, H.; Yang, M.; Xie, D.; Sun, X.; Meng, J.; Wang, Q.; Wu, C. Biodrying of Biogas Residue through a Thermophilic Bacterial Agent Inoculation: Insights into Dewatering Contribution and Microbial Mechanism. *Bioresour. Technol.* **2022**, *355*, 127256. [CrossRef]
30. Ulbricht, M. Advanced Functional Polymer Membranes. *Polymer* **2006**, *47*, 2217–2262. [CrossRef]
31. Gao, Z.; Ma, Y.; Liu, Y.; Wang, Q. Waste Cooking Oil Used as Carbon Source for Microbial Lipid Production: Promoter or Inhibitor. *Environ. Res.* **2022**, *203*, 111881. [CrossRef] [PubMed]
32. Bamforth, C. *Brewing Materials and Processes: A Practical Approach to Beer Excellence*; Academic Press: Cambridge, MA, USA, 2016.
33. Gryta, M.; Tomczak, W. Microfiltration of Post-Fermentation Broth with Backflushing Membrane Cleaning. *Chem. Papers* **2015**, *69*, 544–552. [CrossRef]
34. Yang, Y.; Zhang, J.; Zong, X.; Liu, W. Applied Research of the Ultra Filters Technology in the Salt Water Cheese. *Dairy Ind.* **2011**, *1*, 4.
35. Zhang, Z.; Han, X.; Fang, S.; Chang, C. Study on Ethanol Fermentation Wastewater Treatment by Membrane Separation Method. *Technol. Water Treat.* **2018**, *44*, 57–60.
36. Gul, A.; Hruza, J.; Yalcinkaya, F. Fouling and Chemical Cleaning of Microfiltration Membranes: A Mini-Review. *Polymers* **2021**, *13*, 846. [CrossRef]
37. Petsko, G.; Ringe, D. *Protein Structure and Function*; Oxford University Press: New York, NY, USA, 2003.
38. Mulder, M.H.V. Chapter 2 Polarization Phenomena and Membrane Fouling. In *Membrane Separations Technology*; Noble, R.D., Stern, S.A., Eds.; Membrane Science and Technology; Elsevier: Amsterdam, The Netherlands, 1995; Volume 2, pp. 45–84.
39. Walrant, A.; Vogel, A.; Correia, I.; Lequin, O.; Olausson, B.E.S.; Desbat, B.; Sagan, S.; Alves, I.D. Membrane Interactions of Two Arginine-Rich Peptides with Different Cell Internalization Capacities. *Biochim. Biophys. Acta Biomembr.* **2012**, *1818*, 1755–1763. [CrossRef] [PubMed]
40. Kuberkar, V.T.; Davis, R.H. Microfiltration of Protein-Cell Mixtures with Crossflushing or Backflushing. *J. Membr. Sci.* **2001**, *183*, 1–14. [CrossRef]
41. Hou, L.; Gao, K.; Li, P.; Zhang, X.; Wang, Z.; Song, P.; Yao, W. A Kinetic Model for Calculating Total Membrane Fouling Resistance in Chemical Cleaning Process. *Chem. Eng. Res. Design* **2017**, *128*, 59–72. [CrossRef]
42. Awanis Hashim, N.; Liu, Y.; Li, K. Stability of PVDF Hollow Fibre Membranes in Sodium Hydroxide Aqueous Solution. *Chem. Eng. Sci.* **2011**, *66*, 1565–1575. [CrossRef]
43. Siembida-Lösch, B. Physical Cleaning. In *Encyclopedia of Membranes*; Enrico, D., Giorno, L., Eds.; Springer: Berlin/Heidelberg, Germany, 2015; pp. 1–4. ISBN 978-3-642-40872-4.
44. Luo, J.; Ding, L.; Qi, B.; Jaffrin, M.Y.; Wan, Y. A Two-Stage Ultrafiltration and Nanofiltration Process for Recycling Dairy Wastewater. *Bioresour. Technol.* **2011**, *102*, 7437–7442. [CrossRef]

membranes

Article

Phenylalanine Losses in Neutralization Dialysis: Modeling and Experiment

Anton Kozmai [1,*], Mikhail Porozhnyy [1], Violetta Gil [1] and Lasaad Dammak [2]

[1] Membrane Institute, Kuban State University, 149, Stavropolskaya Str., 350040 Krasnodar, Russia; porozhnyj@mail.ru (M.P.); violetta_gil@mail.ru (V.G.)

[2] Institut de Chimie et des Materiaux Paris-Est (ICMPE), UMR 7182 CNRS—Universite Paris-Est Creteil, 2 Rue Henri Dunant, 94320 Thiais, France; dammak@u-pec.fr

* Correspondence: kozmay@yandex.ru; Tel.: +7-952-86-21-139

Abstract: A non-steady state mathematical model of an amino acid (phenylalanine (Phe)) and mineral salt (NaCl) solution separation by neutralization dialysis (ND) carried out in a batch mode is proposed. The model takes into account the characteristics of membranes (thickness, ion-exchange capacity, and conductivity) and solutions (concentration, composition). As compared to previously developed models, the new one considers the local equilibrium of Phe protolysis reactions in solutions and membranes and the transport of all the phenylalanine forms (zwitterionic, positively and negatively charged) through membranes. A series of experiments on ND demineralization of the NaCl and Phe mixed solution was carried out. In order to minimize Phe losses, the solution pH in the desalination compartment was controlled by changing the concentrations of the solutions in the acid and alkali compartments of the ND cell. The validity of the model was verified by comparison of simulated and experimental time dependencies of solution electrical conductivity and pH, as well as the concentration of Na^+, Cl^- ions, and Phe species in the desalination compartment. Based on the simulation results, the role of Phe transport mechanisms in the losses of this amino acid during ND was discussed. In the experiments carried out, the demineralization rate reached 90%, accompanied by minimal Phe losses of about 16%. Modeling predicts a steep increase in Phe losses when the demineralization rate is higher than 95%. Nevertheless, simulations show that it is possible to achieve a highly demineralized solution (by 99.9%) with Phe losses amounting to 42%.

Keywords: ion-exchange membrane; neutralization dialysis; modeling; phenylalanine losses; amino acid; demineralization

Citation: Kozmai, A.; Porozhnyy, M.; Gil, V.; Dammak, L. Phenylalanine Losses in Neutralization Dialysis: Modeling and Experiment. *Membranes* **2023**, *13*, 506. https://doi.org/10.3390/membranes13050506

Academic Editor: Daqi Cao

Received: 22 April 2023
Revised: 6 May 2023
Accepted: 9 May 2023
Published: 11 May 2023

1. Introduction

In 1986, M. Igawa et al. proposed [1] a new desalination method based on the principle of Donnan dialysis. In a three-compartment cell, the electrolyte solution was separated from the HCl solution using a cation-exchange membrane (CEM) and from the NaOH solution using an anion-exchange membrane (AEM). Cations and anions were exchanged with protons and hydroxide ions through the corresponding membranes, simultaneously providing the processes of desalination and neutralization. Later, the same authors suggested calling this method neutralization dialysis (ND) and demonstrated its applicability for the mutual separation of electrolytes and non-electrolytes in the example of the mixed aqueous solution of potassium chloride and methyl alcohol [2].

The ND method does not require applying an electric field (such as electromembrane methods) and high pressure (such as baromembrane methods). In general, ND is characterized by low energy costs and ease of technical implementation and does not demand high investments, which determines its prospects in terms of application in diverse fields. To date, ND has been found to be used in a wide range of different applications, for example, to neutralize acidified effluents from enterprises in various industries [3], purification of tap water [4], and surface water desalination for producing potable water [5]. In recent years,

ND has been introduced as a basic element of the energy-harvesting technique utilizing waste acid and alkaline solutions (capacitive neutralization dialysis) [6]. Of particular interest is the use of ND for the separation of multicomponent mixtures and the separation of pure valuable components. This method has shown its effectiveness for the separation of weak acids and bases [7], isolation of mono-, oligo- and polysaccharides [8], aldehydes [9], and amino acids [8,10,11]. Another prospect for ND may be its application for the separation of organic and mineral compounds as a pretreatment step for electrodialysis processing of liquids of biological origin, for example, whey. This will allow solving the problem of depressed ion transport due to the presence of organic substances [12,13].

In terms of purification and mutual separation of amino acids, ND has certain advantages over other methods. Amino acids are ampholytes, that is, they have both acidic and basic groups in their structure and can enter the proton-transfer reactions. Such substances are characterized by different values of the equilibrium constants for protonation/deprotonation reactions. The relation of these constants determines the pH values of the medium at which amino acids can change their form (zwitterionic, positively charged, negatively charged). The main advantage of ND is the ability to control the pH value of the processed solution without adding reagents. This allows controlling the form of the ampholytes and, consequently, their fluxes across the membranes. Thus, using ND, it is possible to carry out effective demineralization, concentration, and fractionation of amino acids in the food, pharmaceutical, and biochemical industries, where the introduction of additional reagents can adversely affect the quality of the final product.

However, maintaining the required pH value of the solution being desalted is a demanding task. This value is influenced by a large number of parameters of ion-exchange membranes (for example, the ion-exchange capacity, nature of functional groups, thickness, degree of matrix crosslinking) and of the ND process (concentration and flow rate of acid, alkali solutions, and solution being desalinated in the corresponding compartments) [5,10,11,14,15]. In contrast to ampholyte-containing solutions, the mechanisms of strong electrolyte ions transfer during the ND have been studied sufficiently to gain a comprehensive idea of what trends the pH of the solution being desalted obeys. In addition to a fairly large number of experimental works, a number of mathematical models have been proposed to explain the effects arising in such systems [14,16,17]. It is noted in [5,14] that the pH of a being desalted solution of a strong binary electrolyte can change from acidic to alkaline values (or vice versa, from alkaline to acidic). This occurs due to a change in the kinetics of ion exchange from internal diffusion (transfer is limited by the membrane) to external diffusion (transfer is limited by the diffusion boundary layers), which determines the rate of exchange across the ion-exchange membranes. The study [17] showed that strong pH variations occur back and forth multiple times and are associated with changes in the rate of exchange across the CEM and AEM caused by changes in the concentration of the solution being desalinated. Thus, understanding the mechanisms of the ND process allows for controlling the exchange rate across the ion-exchange membranes and managing the pH values by varying the conditions of the process.

Knowledge about the trends of pH behavior in ND of ampholyte-containing solutions is fragmentary. It is known that pH behaves differently compared to the case of strong electrolytes, which is due to the buffer capacity of ampholytes and their ability to enter into protonation/deprotonation reactions [11,18]. The transport mechanisms in such systems are more complex due to the interaction of a large number of particles, the conjunction of their fluxes, and changes in the form of ampholyte species. Thus, pH control in the case of ampholyte-containing solutions is a difficult task. The selection of optimal conditions for the ND process to minimize the loss of target components through an exclusively experimental approach is rather time, labor, and resource-consuming. For these purposes, the use of mathematical modeling is of great importance. The modeling will assist in deepening the understanding of the transport mechanisms and help to predict the effect of changing one or another process parameter.

An attempt to simulate the process of demineralization of an ampholyte-containing solution (a mixture of phenylalanine and mineral salt) during ND was made in [18]. The model assumes that ampholyte species are not transferred across the membranes. They only can enter into protonation/deprotonation reactions in the desalination compartment due to H^+ and OH^- ions transfer through the CEM and AEM, respectively. This model did not allow for estimating the losses of the ampholyte during solution demineralization, as well as describing the mechanisms of ampholyte species transport across the membranes.

In our previous work [11], we proposed a mathematical model describing the process of neutralization dialysis of the mixed solution of phenylalanine (Phe) and sodium chloride. The model considered local changes in the concentration of mineral salt ions and Phe species (in the form of cation, anion, and zwitterion) in the desalination compartment. Additionally, the transport of charged forms of Phe across the membranes and diffusion boundary layers (DBLs) was taken into account. Theoretically found values of amino acid losses were lower than the experimental ones. This difference can be explained by the fact that the model does not take into account the transfer of the zwitterionic form of Phe across membranes. The membrane's internal pH differs by 1–2 units from that of the external solution [19]. When entering the membrane, the charged forms of Phe appear, implementing the facilitated transport mechanism [20,21].

The main goal of this work is to assess the applicability of ND to the demineralization of a mixed mineral salt and amino acid solution in terms of amino acid losses. For this purpose, we propose a new non-steady state mathematical model to describe the Phe and NaCl mixture demineralization by the ND method in the batch mode. Unlike the previous model [11], the new one takes into account the transfer of not only charged forms but also the zwitterionic form of Phe across the membranes and DBLs. The ability of amino acid to change its electric charge due to protonation/deprotonation reactions and the local changes in the concentration of Na^+, Cl^-, H^+, OH^- ions, and Phe species (in charged and zwitterionic forms) are also considered. The developed model takes into account the main characteristics of membranes (thickness, ion-exchange capacity, and electrical conductivity) and solutions (concentrations, diffusion coefficients of components, and equilibrium constants for protonation/deprotonation reactions), as well as the flow rate of the solutions in the dialyzer compartments. We show that the model adequately describes the experimental data obtained for the equimolar mixture NaCl + Phe in the series of experiments where the initial acid and alkali concentrations were varied. We also demonstrate the ability of the new model to predict the ND process behavior and amino acid losses if experimental data are limited.

2. Theoretical

2.1. System under Study

The system under study consists of three compartments: acid (A), alkali (B), and desalination compartment (D), separated from each other by the CEM and AEM (Figure 1).

Each of the circuits formed by the corresponding compartments together with the elements of the hydraulic system (tubes, intermediate tanks) has volumes V^A, V^B, and V^D for the acid, alkali, and desalination compartments, respectively. Solutions of acid (HCl), alkali (NaOH), and mixed solution (NaCl + Phe) circulate through compartments A, B, and D, respectively, and through intermediate tanks.

Convective transport within the DBLs is neglected. Implicitly, it is taken into account by setting the DBL thickness. It is assumed that all the DBLs have equal thickness. This is justified by the same hydrodynamic conditions in the compartments. It is assumed that the concentrations of all ions do not change along the desalination compartment. In other words, the concentrations of ions at any moment in time are the same both in the volumes of the compartments and in the corresponding intermediate tanks. In the conditions of the experiment described in Section 3, these assumptions are justified by the short length of the cell (2.7 cm) and the pipes (60 cm) used in the hydraulic circuit, as well as a relatively high velocity of solution flowing. The average time of solution passage through the D

compartment (about 2 s) as well as through the overall circuit (about 40 s) is small compared to the duration of a single experimental run (86,000 s). During 40 s, the concentration of the solution in the circuit may change only by 0.02%, which is small compared to the concentration measurement error. Therefore, the material balance equations are applied to the whole volume of solutions circulating through acid, alkali, and desalination circuits.

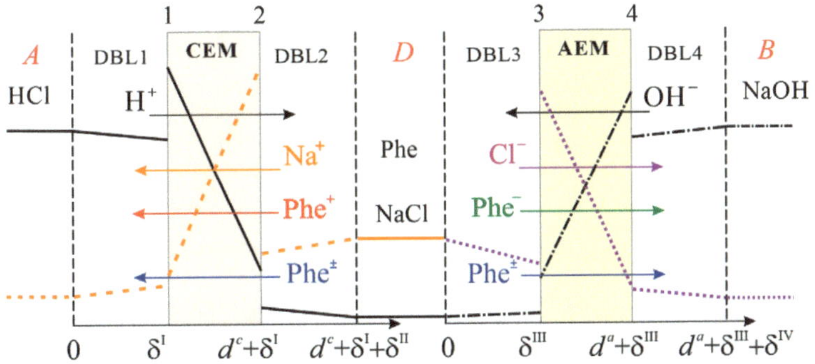

Figure 1. Schematic representation of the modeled system geometry. DBL1 and DBL2 are the diffusion boundary layers adjacent to the CEM from the side of the acid and desalination compartments, respectively; DBL3 and DBL4 are diffusion boundary layers adjacent to the AEM from the side of the desalination and alkali compartments, respectively; indices 1, 2, 3, and 4 denote the boundaries of the CEM and AEM marches with the corresponding compartments.

In contrast to the simplification used in work [18], the concentration of phenylalanine changes in the course of ND in membranes, desalination, acid, and alkali compartments. In contrast to the simplification used in work [11], the local equilibrium of Phe protonation/deprotonation reactions in membranes and solutions, as well as the transport of its charged and zwitterionic forms in membranes and DBLs, are taken into account.

At the membrane/solution interfaces, a local thermodynamic equilibrium between counterions is assumed. CEM and AEM are assumed to be ideally selective (the transport of co-ions is neglected), which is due to the use of relatively dilute solutions in the study.

2.2. Problem Formulation

The following set of equations describes one-dimensional ion electrodiffusion transport in the membranes and the DBLs:

the Nernst-Planck equation

$$J_j = -D_j\left(\frac{\partial C_j}{\partial x} + z_j C_j \frac{F}{RT}\frac{\partial \varphi}{\partial x}\right) \tag{1}$$

the electroneutrality condition

$$\sum_j z_j C_j = \omega Q \tag{2}$$

the condition of zero current flow

$$\sum_j z_j J_j = 0 \tag{3}$$

the equation of material balance

$$\frac{\partial C_j}{\partial t} = -\frac{\partial J_j}{\partial x} \tag{4}$$

where C_j is the concentration, J_j is the flux density, z_j is the charge, D_j is the diffusion coefficient of ion j; Q is the membrane ion-exchange capacity, ω can take the values -1, +1, or 0 for an AEM, a CEM, and a solution, respectively; t is the time, x is the coordinate normal to the membranes surfaces, R, T, and F are the gas constant, temperature, and Faraday constant, respectively. In the D compartment j = H^+, OH^-, Na^+, Cl^-, Phe^+, Phe^-, $Phe^{\pm}$; in the A compartment and in CEM j = H^+, Na^+, Phe^+, $Phe^{\pm}$; in the B compartment and in AEM j = OH^-, Cl^-, Phe^-, $Phe^{\pm}$.

Hereinafter, the zwitterion ($^+NH_3 - CH(CH_2C_6H_5) - COO^-$), the cation ($^+NH_3 - CH(CH_2C_6H_5) - COOH$), and the anion ($NH_2 - CH(CH_2C_6H_5) - COO^-$) of Phe are denoted as $Phe^{\pm}$, Phe^+, and Phe^-, respectively.

The equilibrium between H^+, OH^- ions, and water molecules in solution is described by the equation:

$$K_w = C_H \cdot C_{OH} = 10^{-14}\ \text{mol}^2 \cdot \text{L}^{-2} \tag{5}$$

Phenylalanine enters into the protolysis reactions:

$$Phe^+ + H_2O \leftrightarrow Phe^{\pm} + H_3O^+ \tag{6}$$

$$Phe^{\pm} + H_2O \leftrightarrow Phe^- + H_3O^+ \tag{7}$$

The chemical equilibrium constants of reactions (6), K_1, and (7), K_2, at 25 °C are expressed as follows:

$$K_1 = \frac{[Phe^{\pm}][H_3O^+]}{[Phe^+]} = 6.31 \cdot 10^{-3}\ \text{mol} \cdot \text{L}^{-1} \tag{8}$$

and

$$K_2 = \frac{[Phe^-][H_3O^+]}{[Phe^{\pm}]} = 4.90 \cdot 10^{-10}\ \text{mol} \cdot \text{L}^{-1} \tag{9}$$

respectively [22].

The local change in the concentration of Phe species ($Phe^{\pm}$, Phe^+, and Phe^-) are calculated by Equations (8) and (9) as follows:

$$C_{Phe^+} = \frac{C_{Phe^{tot}}}{\frac{K_1K_2}{C_H^2} + \frac{K_1}{C_H} + 1} \tag{10}$$

$$C_{Phe^-} = \frac{K_1K_2C_{Phe^+}}{C_H^2} \tag{11}$$

$$C_{Phe^{\pm}} = \frac{K_1C_{Phe^+}}{C_H} \tag{12}$$

where $C_{Phe^{tot}} = C_{Phe^{\pm}} + C_{Phe^+} + C_{Phe^-}$ is the total concentration of all Phe species [18,20]. Thus, depending on the local pH value of the solution in the ND system, Phe changes its form due to protolysis reactions (Figure 2).

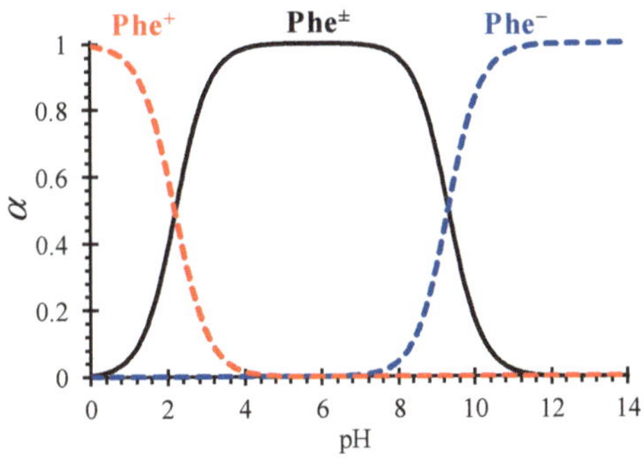

Figure 2. Mole fractions (α) of Phe species in an aqueous solution as function of pH, calculated from Equations (10)–(12).

Changes in ion concentrations in compartments *A*, *D*, and *B*, as well as the local ion concentrations in DBLs and membranes, are calculated from the system of equations formed by the material balance condition, Equation (4), taking into account Equations (6) and (7):

$$\frac{\partial C_H}{\partial t} = -divJ_H + k_1 C_{Phe^+} - k_{-1} C_{Phe^\pm} C_H + k_2 C_{Phe^\pm} - k_{-2} C_{Phe^-} C_H + k_d C_{H_2O} - k_r C_H C_{OH} \tag{13}$$

$$\frac{\partial C_{OH}}{\partial t} = -divJ_{OH} + k_d C_{H_2O} - k_r C_H C_{OH} \tag{14}$$

$$\frac{\partial C_{Phe^+}}{\partial t} = -divJ_{Phe^+} - k_1 C_{Phe^+} + k_{-1} C_{Phe^\pm} C_H \tag{15}$$

$$\frac{\partial C_{Phe^-}}{\partial t} = -divJ_{Phe^-} + k_2 C_{Phe^\pm} - k_{-2} C_{Phe^-} C_H \tag{16}$$

$$\frac{\partial C_{Phe^\pm}}{\partial t} = -divJ_{Phe^\pm} + k_1 C_{Phe^+} - k_{-1} C_{Phe^\pm} C_H - k_2 C_{Phe^\pm} + k_{-2} C_{Phe^-} C_H \tag{17}$$

$$\frac{\partial C_{Na}}{\partial t} = -divJ_{Na} \tag{18}$$

$$\frac{\partial C_{Cl}}{\partial t} = -divJ_{Cl} \tag{19}$$

where k_1, k_2 are the dissociation rate constants, k_{-1}, k_{-2} are the recombination rate constants in reactions described by Equations (6) and (7), respectively; k_d, k_r are the dissociation and recombination rate constants of water.

Solving the system of Equations (1)–(3) and (13)–(19) allows one to find the change in concentrations of H^+, Na^+, Phe^+, $Phe^\pm$ ions in the acid compartment, CEM, and adjacent DBLs (DBL1 and DBL2); OH^-, Cl^-, Phe^-, $Phe^\pm$ ions in the alkali compartment, AEM, and adjacent DBLs (DBL3 and DBL4); H^+, OH^-, Na^+, Cl^-, Phe^+, Phe^-, $Phe^\pm$ ions in the desalination compartment.

The boundary conditions imply the ion exchange equilibrium and the flux continuity condition at the membrane/solution interfaces.

Local equilibrium at the membrane/solution interfaces is described by the equations:

$$K^c_{H,Na} = C^c_H C_{Na} / (C^c_{Na} C_H) \tag{20}$$

$$K^c_{H,Phe^+} = C^c_H C_{Phe^+} / (C^c_{Phe^+} C_H) \tag{21}$$

$$K^a_{OH,Cl} = C^a_{OH} C_{Cl} / (C^a_{Cl} C_{OH}) \tag{22}$$

$$K^a_{OH,Phe^-} = C^a_{OH}C_{Phe^-} / (C^a_{Phe^-}C_{OH}) \tag{23}$$

where C^c_j and C^a_j are the concentration of ion j at the membrane solution interface from the inside of the membrane (superscripts "c" and "a" relates to CEM and AEM, respectively); K^c is the ion exchange equilibrium constant for H^+/Na^+ and H^+/Phe^+ (at the CEM surface from the inside of the membrane); K^a is the similar parameter for OH^-/Cl^- and OH^-/Phe^- (at the AEM surface from the inside of the membrane).

The continuity of the flux condition at the CEM/solution boundaries reads as:

$$J_j\big|_{x=\delta^I} = J^c_j\Big|_{x=\delta^I},\ J_j\big|_{x=d^c+\delta^I+\delta^{II}} = J^c_j\Big|_{x=d^c+\delta^I+\delta^{II}} \tag{24}$$

At the DBLs/solutions boundaries, the concentration continuity condition is set:

$$(C_j)_{x=0} = C^A_j,\ (C_j)_{x=d^c+\delta^I+\delta^{II}} = C^D_j \tag{25}$$

where C^A_j, C^D_j are the concentration of ion j in A and D compartments, respectively.

Similar boundary conditions are set from the AEM side.

The system of the partial differential equations described above was solved numerically using Matlab software.

At the beginning of the ND process (at $t = 0$), a uniform distribution of concentrations in the diffusion layers is assumed, equal to the initial concentrations of the feed solutions. Generally, the initial conditions read as

$$C_j(x)\big|_{t=0} = \begin{cases} C^0_j,\ 0 \le x \le \delta^I \text{ for the DBL1} \\ C^0_j,\ d^c+\delta^I \le x \le d^c+\delta^I+\delta^{II} \text{ for the DBL2} \\ C^0_j,\ 0 \le x \le \delta^{III} \text{ for the DBL3} \\ C^0_j,\ d^a+\delta^{III} \le x \le d^a+\delta^{III}+\delta^{IV} \text{ for the DBL4} \\ C^c_j(x),\ \delta^I \le x \le d^c+\delta^I \text{ for the CEM} \\ C^a_j(x),\ \delta^{III} \le x \le d^a+\delta^{III} \text{ for the AEM} \end{cases} \tag{26}$$

where C^0_j is the initial concentration of species j in the corresponding compartment,$C^c_j(x)$ and $C^a_j(x)$ are linear functions that distribute the concentration of species j linearly between left-hand and right-hand membrane boundaries.

2.3. *Parameters of the Model*

The input parameters of the model can be conventionally divided into three groups: thermodynamic, kinetic, and those that characterize the nature of the ion-exchange material. There are also several input parameters characterizing the solution and the ND system: electrolyte concentrations, pH, and DBLs thickness.

Thermodynamic parameters include ion exchange equilibrium constants between membranes and external solutions and chemical equilibrium constants of Phe protonation/deprotonation reactions. For simplicity, the ion exchange equilibrium constants are taken equal to 1.

Kinetic parameters include the diffusion coefficients of mineral salt and Phe species in membranes and solutions.

The parameter characterizing the nature of the membrane material is the membrane ion-exchange capacity.

The output parameters of the model are the concentrations of ions and Phe species in the membranes, DBLs, and in compartments of the studied system.

The input parameters of the system under study were obtained from independent experiments or taken from the literature. The values of the fitting parameters were found from the condition of the best fit between the simulated and experimental time dependencies of pH, electrical conductivity, and concentrations of mineral salt ions and Phe in the solution in the D compartment. The fitting parameters of the model are the diffusion coefficients of ions and Phe species in membranes ($\overline{D}_j$) and the thickness of DBLs (δ).

Note that the order of magnitude of $\overline{D}_j$ is estimated from experimental data on the membrane electrical conductivity using the Nernst-Einstein relation, as seen in work [17].

3. Experimental

3.1. Neutralization Dialysis Process

Neutralization dialysis was carried out in a three-compartment dialysis cell (similar to one used in [11]) consisting of acid and alkali compartments, as well as a desalination compartment separated by a CEM and an AEM. In each of the compartments, a spacer of the width equal to the intermembrane distance (0.6 cm) was placed. The membrane's working surface area was 7.29 cm^2 (2.7 cm $\times$ 2.7 cm).

The average linear flow rate of the solutions in each compartment was 1.5 $cm \cdot s^{-1}$. A total of 2 L of HCl solution and 2 L of NaOH solution of various concentrations, respectively, circulated through the acid and alkali compartments. A total of 0.5 L of a mixed solution of NaCl and Phe containing dissolved components in an equimolar ratio (0.02 $mol \cdot L^{-1}$ NaCl + 0.02 $mol \cdot L^{-1}$ Phe) was circulated through the desalination compartment.

The initial pH value of the NaCl + Phe mixed solution was 5.5–5.6, which is close to the Phe isoelectric point (pI = 5.76). Such conditions (according to the estimates by Equations (10)–(12)) provide the molar fraction of the zwitterionic form of Phe in the mixed solution in the range from 99.93% to 99.94%.

In the course of neutralization dialysis, the pH and electrical conductivity of the solution being demineralized were measured using an Expert-001 pH meter and an Expert-002 conductometer (JSC Econiks-Expert, Moscow, Russia). The experiments were carried out at a temperature of 25.0 $\pm$ 0.5 °C.

To assess the loss of Phe and the demineralization rate during ND, sampling of the solution in the D compartment was carried out after 12 and 24 h (in the middle and at the end of the experiment). The Phe concentration was determined using the UV-1800 spectrophotometer (TM ECO-VIEW, Shanghai Mapada Instruments Co., Ltd., Shanghai, China) at a wavelength of 259 nm. The content of Na^+ and Cl^- ions was determined using a DIONEX ICS-3000 (Thermo Fisher Scientific, Waltham, MA, USA) chromatographic system.

3.2. Membranes

New commercial homogeneous membranes manufactured by ASTOM (Tokyo, Japan) were used in the ND process: a cation-exchange membrane CSE and an anion-exchange membrane ASE. These are standard-grade, homogeneous membranes with high mechanical strength. The polymer matrix of the CSE membrane is a styrene-based copolymer with $-SO^{3-}$ functional groups; the ASE membrane containing strongly basic functional groups is produced from a copolymer of styrene and divinylbenzene. In the case of both membranes, the reinforcing net is made of a mixture of polyethylene and polypropylene [23].

Some physicochemical characteristics of the CSE and ASE membranes are shown in Table 1.

Table 1. Some physicochemical characteristics of CSE and ASE membranes. Data taken from [11].

Membrane	CSE	ASE
Ion-exchange capacity, $mmol \cdot cm^{-3}$	1.85	1.99
Water content, %	42.0	24.4
Conductivity (in 0.1 M NaCl), $mS \cdot cm^{-1}$	7.37	3.77
Conductivity (in 0.1 M Phe, CSE at pH = 0.5, ASE at pH = 13.1), $mS \cdot cm^{-1}$	5.67	2.32
Conductivity (CSE in 0.1 M HCl, ASE in 0.1 M NaOH), $mS \cdot cm^{-1}$	71.12	6.76
Thickness in swollen state, microns	140	150

4. Results and Discussions

The behavior of the ND system under conditions similar to those applied in the present research (concentrations of the acid and alkali solutions, average linear velocity of the solutions in the compartments, design of the dialyzer) is well studied for the case of a strong binary 1:1 electrolyte [5,14,17].

The peculiarity of the ND system is possible fluctuations of the solution pH in the *D* compartment. These fluctuations may appear as a result of (1) a change in the kinetics of ion transfer from internal to external diffusion (the ion flux is limited by transport through membranes and through DBLs, respectively) with a dilution of the solution in the *D* compartment [14] and (2) a delay in the formation of concentration profiles in membranes and DBLs as a response to changes in the concentration of the solution in the *D* compartment [17].

If an ampholyte (for example, phenylalanine) is present in the *D* compartment, the pH fluctuations become less pronounced due to the buffering effect of Phe (which enters the protonation/deprotonation reactions with H^+ and OH^- ions) [18]. In this case, charged species of Phe (Phe^+ and Phe^-) are formed in the solution and membranes. Phe^+ and Phe^- are transported through the corresponding membranes as counterions, along with the diffusion of the zwitterionic form ($Phe^{\pm}$), leading to the loss of this amino acid. Therefore, pH control of the solution being desalinated is an important task in the implementation of ND of ampholyte-containing solutions.

In order to reduce the loss of Phe, it is necessary to ensure the reagent-free pH control in the *D* compartment so that the pH value is as close as possible to the pI of the amino acid. Based on the literature [8,10,17,18], pH can be controlled by influencing the ion exchange rate (ion fluxes) across the CEM and AEM. This may be realized in several ways:

(1) by the selection of membranes with desired properties (structure, thickness, ion-exchange capacity, nature of fixed ion-exchange groups, etc.);
(2) by changing the hydrodynamic conditions in the compartments of the ND system, which affects the DBL thickness near the membrane surfaces;
(3) by changing the concentration of acid and/or alkali in the corresponding compartments, which affects the concentration gradient between these compartments and the *D* compartment and, as a consequence, the ion fluxes.

In the authors' opinion, the latter way is the simplest from a practical point of view and easiest in terms of implementation. In the present research, in order to verify the proposed model, a series of experiments with the varied concentration of acid and/or alkali in different runs were conducted.

4.1. pH Fluctuations in Desalination Compartment

In general, the pH behavior depends on the exchange rate (ER) ratio across the CEM + adjacent DBLs and AEM + adjacent DBLs [14,17]. The ER across the AEM + DBLs may be significantly higher or significantly lower than that across the CEM (for example, due to the great difference in concentrations in compartments *A* and *B*, membrane thickness, ion-exchange capacity, etc.), leading, respectively, to the situation where the pH in the *D* compartment will have high or low values [14].

The conditions of the ND experiment set in this work refer to the case of comparable values of the ER across the CEM and AEM. Three cases of acid and alkali solution concentrations are considered (Table 2).

Table 2. Initial concentrations in ND system compartments.

Case	C^A_{HCl}, mol·L^{-1}	C^B_{NaOH}, mol·L^{-1}	C^D_{NaCl}, mol·L^{-1}	$C^D_{Phe^{tot}}$, mol·L^{-1}
case 1	0.1	0.1		
case 2	0.05	0.1	0.02	0.02
case 3	0.1	0.2		

As can be seen from Figure 3, the experimental and simulated (using the developed model) time dependencies of pH and electrical conductivity in the *D* compartment are in good agreement in the time range where the experimental data is available. At the same time, Figure 3 shows further pH and electrical conductivity behavior predicted using the developed model. The values of model parameters used in simulations are listed in Table A1, Appendix A.

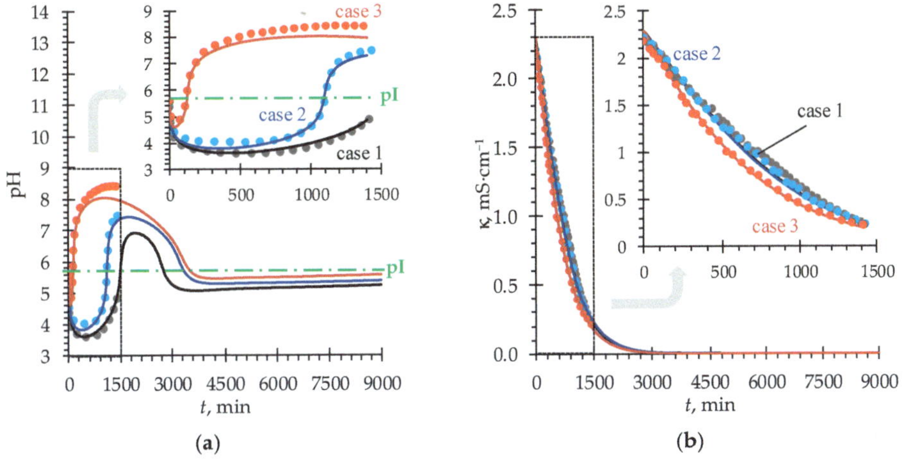

Figure 3. Experimental (dots) and simulated (solid lines) time dependencies of pH (**a**) and electrical conductivity (**b**) of the solution in the *D* compartment.

Figure 3a shows that regardless of the initial values of C^A_{HCl} and C^B_{NaOH} used in our study, the solution pH in the *D* compartment decreases at the beginning of the process (in all three cases). At a sufficiently high concentration of the solution being desalted at the beginning of the ND process, the limiting stage is ion transfer across the CEM and AEM (internal diffusion kinetics) and not the DBLs facing the *D* compartment. The mutual diffusion coefficient of H^+/Na^+ ions in the CEM exceeds the corresponding value for the OH^-/Cl^- ions in the AEM (calculated similarly as in [5]), which is expressed in larger values of the fluxes through the corresponding membranes of the former compared to the latter (Figure 4).

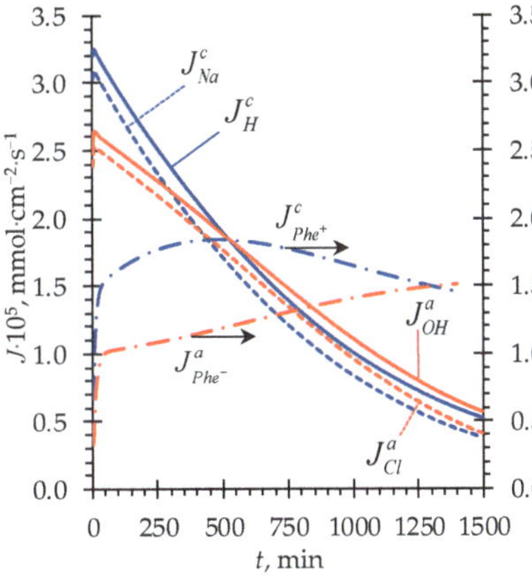

Figure 4. Simulated time dependencies of ion fluxes through the CEM (J^c_j, j = H^+, Na^+, Phe^+) and AEM (J^a_j, j = OH^-, Cl^-, Phe^-). Case 1 is considered (Table 2).

Further, in the course of ND, the pH behavior in the D compartment follows different trends for each of the considered cases up to approximately 2000 min. Let us consider these differences.

In case 1, as the concentration of the solution in the D compartment decreases, the role of the DBLs (where the mutual diffusion coefficient for the OH^-/Cl^- ions is higher than for the H^+/Na^+ ions) in controlling the transport kinetics increases and becomes comparable to the role of the membranes. Thus, the ion fluxes through the AEM + adjacent DBLs and CEM + adjacent DBLs are equalized (at $t \approx 500$ min, Figure 4), which results in an almost constant pH value (Figure 3a). Later, the concentration of the solution decreases to a threshold value [14], after reaching which the process begins to be controlled by external diffusion kinetics since the transfer in the DBLs becomes the limiting stage. The mutual diffusion coefficient of OH^-/Cl^- ions in solution is approximately 20% greater than that of H^+/Na^+ ions [5,17], therefore, there is a trend towards an increase in pH.

In case 2, using a lower initial acid concentration leads to a decrease in H^+ ion fluxes through the CEM as compared to case 1. As a result, at the beginning of the process, the deviation of the pH value from the initial one turned out to be less pronounced (Figure 3a). As a consequence, when the external diffusion kinetics takes over the main role in controlling the process, the increase in pH is more prominent and occurs earlier than in case 1.

In case 3, the initial concentration of alkali was doubled compared to case 1. The ER across the AEM quickly becomes the determining factor of the process, which is expressed in a sharp increase in pH in the D compartment even before reaching the threshold concentration at which the external diffusion kinetics begins to dominate. Apparently, this behavior of the system is due to a significant difference in the concentration gradient of OH^- ions between the B and D compartments and the concentration gradient of H^+ ions between the A and D compartments.

After approximately 2500 min of the process, the model predicts one more shift in pH toward acidic values in all three considered cases. By this time, due to a greater ER across the AEM + adjacent DBLs, the concentration of Cl^- ions becomes lower than that of Na^+ ions (Figure 5). This makes the ER across the CEM + adjacent DBLs greater again and causes the pH in the D compartment to decrease (Figure 3a). As a result, after about 3000 min of the process, the concentration of Na^+ ions becomes lower than that of Cl^- ions (Figure 5), and the balance of ER between ion-exchange membranes shifts in favor of the AEM + adjacent DBLs. However, by this time, the concentration of mineral salt ions becomes very low (of the order of the Phe charged forms concentrations), yielding very low values of ion fluxes. Thus, the further changes in pH become very time-expanded. In Figure 3a, starting from approximately 3750 min, the pH time dependence has a shape of a straight line. When the concentration of mineral salt tends to zero, the system tends to an equilibrium state described by Equations (8) and (9), and pH in the D compartment tends to the value of pI of Phe. Further, the transport of H^+ and OH^- ions into the D compartment will stop since these ions will be spent on local protonation/deprotonation of the zwitterionic form of Phe in membranes, followed by transfer of charged Phe^+ and Phe^- species towards A and B compartments from the CEM and AEM, respectively. The limiting stage of amino acid transport will be its diffusion through the DBLs to the membrane surfaces facing the D compartment.

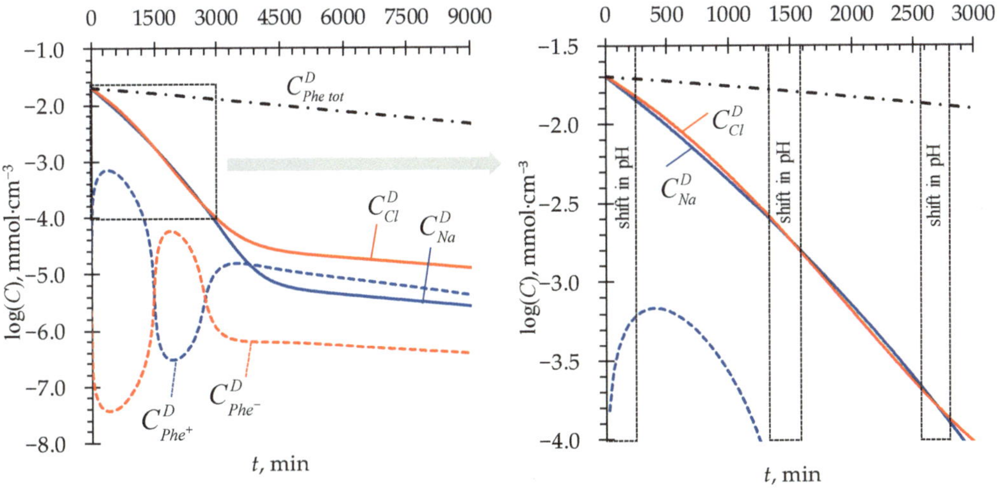

Figure 5. Simulated time dependencies of Na^+, Cl^- (solid lines), Phe^+, Phe^- (dashed lines), and Phe^{tot} (dash-dotted line) concentrations in the *D* compartment. Case 1 is considered (Table 2).

4.2. Phenylalanine Losses

One of the goals of this study is to estimate the Phe losses and to elucidate the mechanisms standing behind the Phe transport in the studied ND system.

The Phe transport is carried out by the diffusion of its charged (Phe^+, Phe^-) and zwitterionic ($Phe^\pm$) forms through the membranes and DBLs.

If the solution pH in the *D* compartment deviates from pI of Phe, the protonation/deprotonation of Phe occurs to form its charged forms that are transferred through the membranes as counterions: Phe^+ through the CEM and Phe^- through the AEM. An additional mechanism of Phe transport is so-called facilitated diffusion [24]. It is known from the literature that the pH in the ion-exchange membrane differs from that in the equilibrium solution by 1–2 units [19] (it is lower in the CEM and higher in the AEM due to the Donnan effect [25]). Entering the membrane, the Phe zwitterion becomes protonated (in CEM) or deprotonated (in AEM) to form additional Phe^+ or Phe^- ions, respectively. This provides an increase in the concentration of charged forms of Phe in the corresponding membranes and intensifies the Phe flux. The limiting factor of such a transport mechanism is the $Phe^\pm$ diffusion rate through the DBLs towards the membranes in the *D* compartment. Thus, apparently, the loss of Phe associated with the transport of its zwitterionic form cannot be avoided by controlling the pH of the solution in the desalination compartment.

However, one may try to minimize the loss of Phe transported in the form of charged ions that appeared in the *D* compartment as a result of the change in pH caused by the counter fluxes of H^+ and OH^- ions.

In the developed model, accounting for the chemical equilibrium of Phe as a function of local pH in solutions and membranes (Equations (8) and (9)) made it possible to describe adequately the experimentally determined concentrations of phenylalanine and mineral salt ions in the solution being desalted (Figure 6).

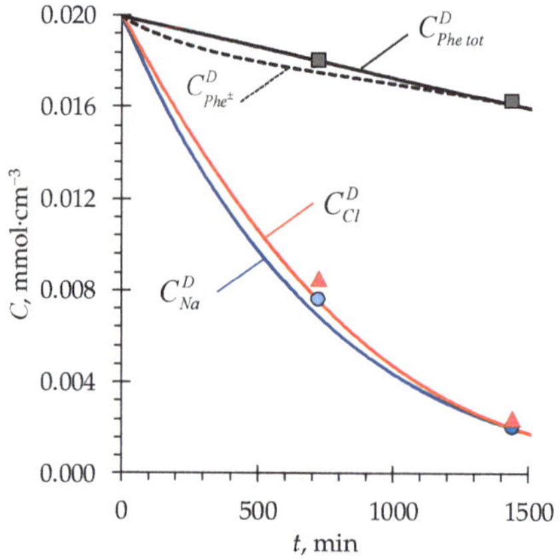

Figure 6. Experimental (markers) and simulated (lines) time dependencies of Na^+, Cl^-, $Phe^\pm$, and Phe^{tot} concentrations in the *D* compartment. Case 1 is considered (Table 2).

Knowing the Phe concentrations, the loss of amino acid was calculated by the equation:

$$L = 100\left(C^D_{Phe\ tot,0} - C^D_{Phe\ tot,t}\right)/C^D_{Phe\ tot,0} \quad (27)$$

where $C^D_{Phe\ tot,0}$ is the initial concentration of Phe in the *D* compartment (at *t* = 0), $C^D_{Phe\ tot,t}$ is the sum of the Phe concentrations in all its forms (Phe^+, Phe^-, $Phe^\pm$) in the *D* compartment at a given moment in time (*t*).

Figure 7 shows that the least loss of Phe is in case 2 (among the three considered cases). In this case, the minimal amount (among all the cases) of Phe^+ and Phe^- ions transferred from the D compartment is conditioned not only by the magnitude of pH deviations from the value of pI but also by the duration of such deviations (Figure 3a).

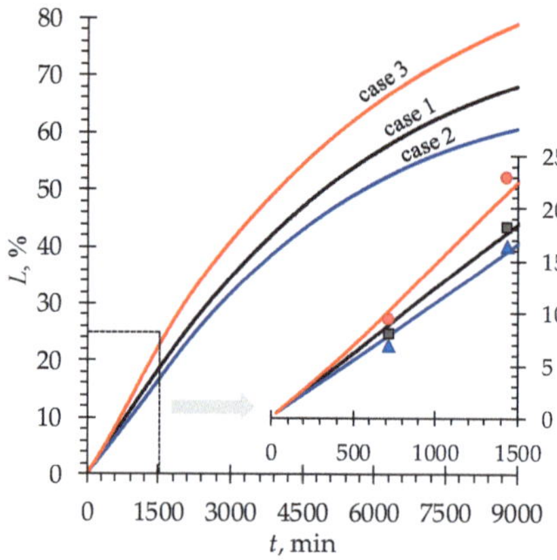

Figure 7. Simulated (lines) and experimentally determined (markers) time dependencies of Phe^{tot} loss in the *D* compartment for the three considered cases (Table 2).

The question of practical importance is what Phe losses can be expected in the ND of an equimolar NaCl + Phe solution depending on the demineralization rate, calculated by the formula:

$$DR = 100\left(C^D_{NaCl,0} - C^D_{NaCl,t}\right)/C^D_{NaCl,0} \quad (28)$$

where $C^D_{NaCl,0}$ is the initial concentration of NaCl in the *D* compartment (at t = 0), $C^D_{NaCl,t}$ is the NaCl concentration in the *D* compartment at a given moment in time (t).

In the ND experiment, 90% DR was achieved, accompanied by minimal Phe losses of about 16% (case 2). Modeling predicts a steep increase in phenylalanine losses when the DR is higher than 95% (Figure 8). At such negligibly small concentrations of mineral salt ions, the ER across the membranes is very low. Thus, a lot of time is required to achieve even an insignificant increase in the DR. During this time, the diffusion of $Phe^{\pm}$ zwitterions leads to an increase in phenylalanine losses. Nevertheless, simulations show that it is possible to achieve a highly demineralized solution characterized by a DR of 99.9% with phenylalanine losses of about 42%.

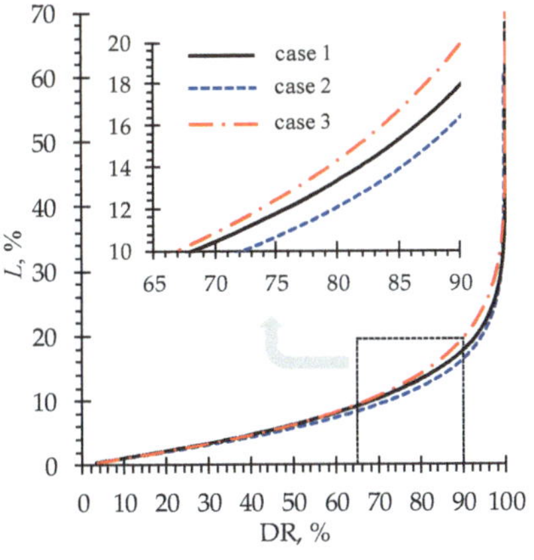

Figure 8. Simulated dependencies of Phe loss as a function of demineralization rate of the solution in the *D* compartment for the three considered cases (Table 2).

Thus, this study has shown that ND can be used to effectively demineralize an amino acid solution, and the control of target product losses is possible by selecting process conditions. The developed model allowed a deeper understanding of the mechanisms that determine the behavior of the ND system, which will permit the ability to deliberately set the optimal parameters. Therefore, it will be possible to increase the attractiveness of ND and simplify the task of controlling the pH during the demineralization of amino acid solutions in comparison with other methods. For example, during electrodialysis demineralization of amino acid solutions, reagent-free pH control is possible by controlling the water splitting process (which can occur at interface boundaries in over-limiting current modes [26]). This is possible by regulating the applied electric current density [27] or by selecting different types of membranes, for example, bipolar ones [28]. However, the problem of water splitting control is rather sophisticated, and this approach requires the application of high voltages [29], thereby, adding chemical reagents is still a common practice used to adjust solution pH in electrodialysis [30]. As for diffusion dialysis, which is also widely used for the demineralization of ampholyte-containing solutions, the control of the pH is possible only with the use of additional reagents. Moreover, the values of fluxes that can be achieved in diffusion dialysis are very low [31], thus, a great number of membranes are required for the demineralization of solutions [32]. The advantage of ND is the relatively high rate of solution desalination. Due to the neutralization reaction in the *D* compartment, a high concentration gradient of H^+ and OH^- ions is maintained, resulting in a relatively high magnitude of ion fluxes through membranes (comparable to ion fluxes in electrodialysis, $\approx 10^{-5}$ mmol·cm^{-2}·s^{-1} [29,33]). This allows for achieving relatively high performance with relatively small plant sizes.

5. Conclusions

A non-steady state model that describes the neutralization dialysis demineralization of a mixed sodium chloride and phenylalanine solution was proposed. The model takes into account the characteristics of membranes that are important for practice (thickness, ion-exchange capacity, conductivity). In contrast to the previously developed models, the new one considers the local equilibrium of phenylalanine protolysis reactions in solutions and membranes and the phenylalanine transport in its charged and zwitterionic forms through membranes.

A series of experiments on neutralization dialysis demineralization of the NaCl and phenylalanine mixed solution was carried out. In order to minimize phenylalanine losses, the solution pH in the desalination compartment was controlled by changing the concentrations of the solutions in the acid and alkali compartments of the dialysis cell.

The validity of the model was verified by comparison of simulated and experimental time dependencies of solution electrical conductivity and pH, as well as the concentration of mineral salt ions and phenylalanine in the desalination compartment of the neutralization dialysis cell.

The results of simulations show that the phenylalanine losses associated with the facilitated diffusion transport mechanism can not be avoided in neutralization dialysis due to the H^+ and OH^- ions fluxes directed oppositely to the fluxes of phenylalanine zwitterions. However, it is possible to lower the transport (consequently, the losses) of charged forms of phenylalanine that formed in the desalination compartment due to pH changes. This reduction can be achieved by regulating the pH in the desalination compartment in a way to keep it as close as possible to the value of amino acid pI. In this study, pH was regulated by influencing the fluxes of H^+ and OH^- ions through the membranes by changing the solution concentrations in the acid and alkali compartments of the cell.

In the experiments carried out, the demineralization rate has reached 90%, accompanied by the minimal phenylalanine losses of about 16%. In the studied system, this result was achieved by using the acid concentration half as much as the alkali concentration. Modeling predicts a steep increase in phenylalanine losses when the demineralization rate is higher than 95%. Nevertheless, simulations show that it is possible to achieve a highly demineralized solution (by 99.9%) with phenylalanine losses amounting to 42%.

Author Contributions: A.K.: Methodology; Research design and investigation; Data analysis and curation; Manuscript writing. M.P.: Data analysis and curation; Project administration, Draft preparation, Funding acquisition. V.G.: Formal Analysis, Investigation, Revision and Validation. L.D.: Revision and Validation. All authors have read and agreed to the published version of the manuscript.

Funding: This work was supported by the Russian Science Foundation (project no. 21-79-00114, https://rscf.ru/en/project/21-79-00114/ (accessed on 10 May 2023)).

Institutional Review Board Statement: Not applicable.

Data Availability Statement: Not applicable.

Acknowledgments: The authors thank the Center for Collective Use of the Kuban State University. "Diagnostics of the structure and properties of nanomaterials" for the equipment provided.

Conflicts of Interest: The authors declare no conflict of interest.

Appendix A

Table A1. Parameters of the model.

Parameter	Value	Units	Description
d^c and d^a	140 and 150	microns	CEM and AEM thickness, respectively *
δ^k	85		The thickness of diffusion boundary layers **, k = *I*, *II*, *III*, *IV* (Figure 1)
C^D_{NaCl}	0.02	mol·L^{-1}	Initial molar concentration of salt in the desalination compartment *D* (Table 2)
C^A_{HCl}	0.1 or 0.05		Initial molar concentration of acid in the acid compartment *A* (Table 2)
C^B_{NaOH}	0.1 or 0.2		Initial molar concentration of alkali in the alkali compartment *B* (Table 2)
$C^D_{Phe^{tot}}$	0.02		Initial molar concentration of phenylalanine in the desalination compartment *D* (Table 2)
V^A and V^B	2	L	Volume of solutions in the acid and alkali compartments, respectively
V^D	0.5		Volume of the solution in the desalination compartment
Q^c and Q^a	1.85 and 1.99	mmol·cm^{-3}	Ion-exchange capacity of CEM and AEM, respectively (in the swollen state) *
D_{Na}	$1.34 \cdot 10^{-5}$	cm^2·s^{-1}	Diffusion coefficients of Na^+, Cl^-, H^+, OH^-, Phe^+, Phe^-, and $Phe^\pm$ ions in solution at infinite dilution [22,34]
D_{Cl}	$2.03 \cdot 10^{-5}$		
D_H	$9.31 \cdot 10^{-5}$		
D_{OH}	$5.26 \cdot 10^{-5}$		
D_{Phe^+}	$6.7 \cdot 10^{-7}$		
D_{Phe^-}	$6.7 \cdot 10^{-7}$		
$D_{Phe^\pm}$	$6.7 \cdot 10^{-7}$		
$\overline{D}^c_{Na}$	$3.25 \cdot 10^{-7}$		Diffusion coefficients of Na^+, H^+, Phe^+ ions in CEM and Cl_-, OH^-, Phe^- ions in AEM **
$\overline{D}^c_H$	$9.0 \cdot 10^{-6}$		
$\overline{D}^c_{Phe^+}$	$1.1 \cdot 10^{-7}$		
$\overline{D}^a_{Cl}$	$2.25 \cdot 10^{-7}$		
$\overline{D}^a_{OH}$	$7.8 \cdot 10^{-7}$		
$\overline{D}^a_{Phe^-}$	$0.75 \cdot 10^{-7}$		
$\overline{D}^c_{Phe^\pm}$ and $\overline{D}^a_{Phe^\pm}$	$3.3 \cdot 10^{-8}$ and $2.3 \cdot 10^{-8}$		Diffusion coefficients of $Phe^\pm$ ions in CEM and AEM, respectively **
K_w	10^{-14}	mol^2·L^{-2}	Water ionization constant
K_1	$6.31 \cdot 10^{-3}$	mol·L^{-1}	Equilibrium constant for chemical reaction (6) [22]
K_2	$4.90 \cdot 10^{-10}$		Equilibrium constant for chemical reaction (7) [22]
$K^c_{H,Na}$ and K^c_{H,Phe^+}	1	–	Nikolskii equilibrium constant for exchange between H_+ and Na^+, H^+, and Phe^+ in CEM
$K^a_{OH,Cl}$ and K^a_{OH,Phe^-}	1	–	Nikolskii equilibrium constant for exchange between OH^- and Cl^-, OH^- and Phe^- in AEM
T	298	K	Absolute temperature
R	8.314	J·(mol·K)$^{-1}$	Gas constant
F	96,485	C·mol^{-1}	Faraday constant

* Parameters obtained from independent experiments (Table 1, data taken from [11]). ** Fitting parameters of the model.

Table A2. List of abbreviations and superscripts/subscripts.

Abbreviations	Description
AEM	anion-exchange membrane
CEM	cation-exchange membrane
DBL	diffusion boundary layer
DR	desalination rate
ER	exchange rate
ND	neutralization dialysis
Phe	phenylalanine
Superscripts/subscripts	
Superscripts "*c*" and "*a*"	relates to CEM and AEM, respectively
Superscripts "*A*", "*B*", and "*D*"	relates to acid, alkali (base), and desalination compartments, respectively
Subscript "*j*"	relates to a certain species. The type of species described by the index *j* is indicated in the place of its appearance

References

1. Igawa, M.; Echizenya, K.; Hayashita, T.; Seno, M. Donnan dialysis desalination. *Chem. Lett.* **1986**, *15*, 237–238. [CrossRef]
2. Igawa, M.; Echizenya, K.; Hayashita, T.; Seno, M. Neutralization Dialysis for Deionization. *Bull. Chem. Soc. Jpn.* **1987**, *60*, 381–383. [CrossRef]
3. German, M.; SenGupta, A.K.; Greenleaf, J. Hydrogen ion (H^+) in waste acid as a driver for environmentally sustainable processes: Opportunities and challenges. *Environ. Sci. Technol.* **2013**, *47*, 2145–2150. [CrossRef]
4. Igawa, M.; Mikami, K.; Okochi, H. Transport characteristics of neutralization dialysis and desalination of tap water. *Bull. Chem. Soc. Jpn.* **2003**, *76*, 437–441. [CrossRef]
5. Chérif, M.; Mkacher, I.; Dammak, L.; Ben Salah, A.; Walha, K.; Grande, D.; Nikonenko, V. Water desalination by neutralization dialysis with ion-exchange membranes: Flow rate and acid/alkali concentration effects. *Desalination* **2015**, *361*, 13–24. [CrossRef]
6. Liu, Y.; Zhang, Y.; Ou-Yang, W.; Bastos Sales, B.; Sun, Z.; Liu, F.; Zhao, R. Capacitive neutralization dialysis for direct energy generation. *Environ. Sci. Technol.* **2017**, *51*, 9363–9370. [CrossRef] [PubMed]
7. Igawa, M.; Tanabe, H.; Ida, T.; Yamamoto, F.; Okochi, H. Separation of weak acids and bases by neutralization dialysis. *Chem. Lett.* **1993**, *22*, 1591–1594. [CrossRef]
8. Bleha, M.; Tishchenko, G.A. Neutralization dialysis for desalination. *J. Membr. Sci.* **1992**, *73*, 305–311. [CrossRef]
9. Igawa, M.; Fukushi, Y.; Hayashita, T.; Hoffmann, M.R. Selective transport of aldehydes across an anion-exchange membrane via the formation of bisulfite adducts. *Ind. Eng. Chem. Res.* **1990**, *29*, 857–861. [CrossRef]
10. Vasil'eva, V.I.; Saud, A.M.; Akberova, E.M. Effect of the mass fraction of ion-exchange resin in a Ralex CM cation-exchange membrane on demineralization of phenylalanine aqueous salt solutions by neutralization dialysis. *Membr. Membr. Technol.* **2021**, *3*, 98–106. [CrossRef]
11. Porozhnyy, M.V.; Kozmai, A.E.; Mareev, A.A.; Gil, V.V. Theoretical and experimental study of neutralization dialysis of phenylalanine–mineral salt equimolar mixture of different concentrations. *Membr. Membr. Technol.* **2022**, *4*, 306–318. [CrossRef]
12. Myronchuk, V.; Zmievskii, Y.; Dzyazko, Y.; Rozhdestvenska, L.; Zakharov, V. Whey desalination using polymer and inorganic membranes: Operation conditions. *Acta Period. Technol.* **2018**, *49*, 103–115. [CrossRef]
13. Chen, X.; Chen, G.Q.; Wang, Q.; Xu, T.; Kentish, S.E. Transforming salty whey into cleaning chemicals using electrodialysis with bipolar membranes. *Desalination* **2020**, *492*, 114598. [CrossRef]
14. Denisov, G.A.; Tishchenko, G.; Bleha, M.; Shataeva, L. Theoretical analysis of neutralization dialysis in the three-compartment membrane cell. *J. Membr. Sci.* **1995**, *98*, 13–25. [CrossRef]
15. Saud, A.M.; Vasil'eva, V.I.; Goleva, E.A.; Akberova, E.M.; Kozlov, A.T. Selection of anion-exchange membrane for separation of phenylalanine and sodium chloride by neutralization dialysis. *Sorbtsionnye i Khromatograficheskie Protsessy* **2020**, *20*, 749–759. [CrossRef]
16. Sato, K.; Yonemoto, T.; Tadaki, T. Modeling of ionic transport in neutralization dialytic deionization. *J. Chem. Eng. Jpn.* **1993**, *26*, 68–74. [CrossRef]
17. Kozmai, A.; Chérif, M.; Dammak, L.; Bdiri, M.; Larchet, C.; Nikonenko, V. Modelling non-stationary ion transfer in neutralization dialysis. *J. Membr. Sci.* **2017**, *540*, 60–70. [CrossRef]
18. Kozmai, A.; Goleva, E.; Vasil'eva, V.; Nikonenko, V.; Pismenskaya, N. Neutralization dialysis for phenylalanine and mineral salt separation. Simple theory and experiment. *Membranes* **2019**, *9*, 171. [CrossRef]

19. Franck-Lacaze, L.; Sistat, P.; Huguet, P. Determination of the pK_a of poly (4-vinylpyridine)-based weak anion exchange membranes for the investigation of the side proton leakage. *J. Membr. Sci.* **2009**, *326*, 650–658. [CrossRef]
20. Metayer, M.; Legras, M.; Grigorchouk, O.; Nikonenko, V.; Langevin, D.; Labbe, M.; Lebrun, L.; Shaposhnik, V. Facilitated transport of a-alanine and phenylalanine through sulfonic cation-exchange membranes. *Desalination* **2002**, *147*, 375–380. [CrossRef]
21. Shaposhnik, V.A.; Eliseeva, T.V. Barrier effect during the electrodialysis of ampholytes. *J. Membr. Sci.* **1999**, *161*, 223–228. [CrossRef]
22. Lide, D.R. *Handbook of Chemistry and Physics*; CRC Press: Boca Raton, FL, USA, 2005; ISBN 0849304873.
23. Chen, G.Q.; Wei, K.; Hassanvand, A.; Freeman, B.D.; Kentish, S.E. Single and binary ion sorption equilibria of monovalent and divalent ions in commercial ion exchange membranes. *Water Res.* **2020**, *175*, 115681. [CrossRef] [PubMed]
24. Vasil'eva, V.I.; Shaposhnik, V.A.; Zemlyanukhina, I.A.; Grigorchuk, O.V. Facilitated diffusion of amino acids in ion-exchange membranes. *Russ. J. Phys. Chem. A* **2003**, *77*, 1017–1019.
25. Donnan, F.G. Theory of membrane equilibria and membrane potentials in the presence of non-dialysing electrolytes. *J. Membr. Sci.* **1995**, *100*, 45–55. [CrossRef]
26. Nikonenko, V.V.; Pismenskaya, N.D.; Belova, E.I.; Sistat, P.; Huguet, P.; Pourcelly, G.; Larchet, C. Intensive current transfer in membrane systems: Modelling, mechanisms and application in electrodialysis. *Adv. Colloid Interface Sci.* **2010**, *160*, 101–123. [CrossRef] [PubMed]
27. Eliseeva, T.; Kharina, A. Desalination of neutral amino acid solutions in an electromembrane system. *Membranes* **2022**, *12*, 665. [CrossRef]
28. Lin, X.; Pan, J.; Zhou, M.; Xu, Y.; Lin, J.; Shen, J.; Van der Bruggen, B. Extraction of amphoteric amino acid by bipolar membrane electrodialysis: Methionine acid as a case study. *Ind. Eng. Chem. Res.* **2016**, *55*, 2813–2820. [CrossRef]
29. Nikonenko, V.; Urtenov, M.; Mareev, S.; Pourcelly, G. Mathematical modeling of the effect of water splitting on ion transfer in the depleted diffusion layer near an ion-exchange membrane. *Membranes* **2020**, *10*, 22. [CrossRef]
30. Kattan Readi, O.M.; Gironès, M.; Wiratha, W.; Nijmeijer, K. On the isolation of single basic amino acids with electrodialysis for the production of biobased chemicals. *Ind. Eng. Chem. Res.* **2013**, *52*, 1069–1078. [CrossRef]
31. Luo, J.; Wu, C.; Xu, T.; Wu, Y. Diffusion dialysis-concept, principle and applications. *J. Memb. Sci.* **2011**, *366*, 1–16. [CrossRef]
32. Jiang, S.; Sun, H.; Wang, H.; Ladewig, B.P.; Yao, Z. A comprehensive review on the synthesis and applications of ion exchange membranes. *Chemosphere* **2021**, *282*, 130817. [CrossRef] [PubMed]
33. Elisseeva, T.V.; Shaposhnik, V.A.; Luschik, I.G. Demineralization and separation of amino acids by electrodialysis with ion-exchange membranes. *Desalination* **2002**, *149*, 405–409. [CrossRef]
34. Robinson, R.A.; Stokes, R.H. *Electrolyte Solutions*, 2nd ed.; Dover Publications Inc.: Mignola, NY, USA, 2002; ISBN 0486422259.

membranes

MDPI

Article

Evaluation of Efficiently Removing Secondary Effluent Organic Matters (EfOM) by Al-Based Coagulant for Wastewater Recycling: A Case Study with an Industrial-Scale Food-Processing Wastewater Treatment Plant

Yu Cheng [1], Qiangqiang Cheng [1], Chengjin Zhao [1], Xianghao Ren [1,*], Yu Wang [1], Yingying Kou [1], Kangmin Chon [2,3], Myung-Han Ko [4] and Moon-Hyun Hwang [5,*]

1 Key Laboratory of Urban Stormwater System and Water Environment, Ministry of Education, Beijing University of Civil Engineering and Architecture, Beijing 100044, China
2 Department of Environmental Engineering, College of Engineering, Kangwon National University, 1 Kangwondaehak-gil, Chuncheon-si 24341, Gangwon-do, Republic of Korea
3 Department of Integrated Energy and Infra System, Kangwon National University, 1 Kangwondaehak-gil, Chuncheon-si 24341, Gangwon-do, Republic of Korea
4 ANT21, 34 Gyebaek-ro, Jung-gu, Daejeon 34899, Republic of Korea
5 Institute of Conversions Science, Korea University, 145 Anam-ro, Sungbuk-gu, Seoul 02841, Republic of Korea
* Correspondence: renxianghao@bucea.edu.cn (X.R.); h2sm2000@korea.ac.kr (M.-H.H.)

Abstract: The reuse of wastewater has been identified as an important initiative for the sustainable development of the environment; thus, the removal of secondary effluent organic matter (EfOM) to ensure the safety of reused wastewater is the key step and a subject of extensive research. In this study, $Al_2(SO_4)_3$ and anionic polyacrylamide were selected as coagulant and flocculant, respectively, for the treatment of secondary effluent from a food-processing industry wastewater treatment plant to meet the standard regulatory specifications for water reuse. In this process, the removal efficiencies of chemical oxygen demand (COD), components with UV_{254}, and specific ultraviolet absorbance (SUVA) were 44.61%, 25.13%, and 9.13%, respectively, with an associated reduction in chroma and turbidity. The fluorescence intensities (Fmax) of two humic-like components were reduced during coagulation, and microbial humic-like components of EfOM had a better removal efficiency because of a higher Log Km value of 4.12. Fourier transform infrared spectroscopy showed that $Al_2(SO_4)_3$ could remove the protein fraction of the soluble microbial products (SMP) of EfOM by forming a loose SMP protein complex with enhanced hydrophobicity. Furthermore, flocculation reduced the aromaticity of secondary effluent. The cost of the proposed secondary effluent treatment was 0.034 CNY t^{-1} $\%COD^{-1}$. These results demonstrate that the process is efficient and economically viable for EfOM removal to realize food-processing wastewater reuse.

Keywords: organic matters; secondary effluent; soluble microbial products; coagulation; humic-like substance; water reuse

Citation: Cheng, Y.; Cheng, Q.; Zhao, C.; Ren, X.; Wang, Y.; Kou, Y.; Chon, K.; Ko, M.-H.; Hwang, M.-H. Evaluation of Efficiently Removing Secondary Effluent Organic Matters (EfOM) by Al-Based Coagulant for Wastewater Recycling: A Case Study with an Industrial-Scale Food-Processing Wastewater Treatment Plant. *Membranes* **2023**, *13*, 510. https://doi.org/10.3390/membranes13050510

Academic Editor: Gaetano Di Bella

Received: 26 March 2023
Revised: 13 April 2023
Accepted: 18 April 2023
Published: 12 May 2023

1. Introduction

Biological treatment processes are one of the most widely used technologies for wastewater treatment worldwide due to their economic and sustainable characteristics [1]. Although a major fraction of the organic matter in wastewater is degraded during biotreatment, there is a considerable amount of residual organic matter in the secondary effluents identified as secondary effluent organic matter (EfOM). Conventional activated sludge systems (e.g., anaerobic and oxic processes) have not only the insufficient ability to biodegrade the refractory organic compounds (e.g., humic- and fulvic acid-like substances) but also produce soluble microbial products (SMP) which inevitably end up as EfOM and affect the effluent quality [2]. A high concentration of EfOM adversely affects the reuse of wastewater, and the presence of organic matter in the circulating water has become the main factor restricting wastewater reuse [3]. In particular, the presence of organic substances such as humic acid makes the water treatment process more difficult [4]. For the widespread implementation of industrial wastewater treatment and reuse, the government has set strict water reuse standards to force various industries to reduce EfOM concentrations, e.g., when reclaimed water is used as a source of industrial water, the chemical oxygen demand (COD) must be below 60 mg L^{-1} (GB/T 19923-2005, national standard). In addition, in response to the needs of regional development, the application of reclaimed water has been repeatedly written into the construction plan. The recently enacted "Guidance on Promoting Wastewater Resource Utilization (2021)" states that the utilization rate of reclaimed water in the Beijing–Tianjin–Hebei region (China) is specified to increase to over 35% by 2025 [5]. These limits have prompted investigations to eliminate the EfOM from secondary effluents in effective, economical, and engineering-friendly ways to ensure environmental protection and safety of water use.

Coagulation, a physical–chemical process, can neutralize surface electrical charges on particles, reduce electrical double-layer repulsion forces of colloids, and also realize adsorption and complexation with several organic substances [6,7]. Through these mechanisms, many kinds of particulate, colloidal, and dissolved organic matter from EfOM can be removed during coagulation, which is an important opportunity to increase the production capacity of reclaimed water. The EfOM comprises two kinds of organic components, i.e., humic substances and SMPs, depending on their origin [8]. Humic substances in the EfOM range from 17% to 71%, which not only enhance the transportation of heavy metals but also cause harmful byproduct formation during disinfection [1]; SMPs are the major soluble organic matter constituents in EfOM [8]. Coagulation treatment of secondary effluent to remove these two types of organics is effective, and its potential has been proven: Wang et al. [1] concluded that the coagulation process could remove almost half of the humic substances in EfOM; Liu et al. [9] used $Al_2(SO_4)_3$ and $FeCl_3$ as coagulants and obtained SMP removal rates of approximately 25% in both cases. However, the removal efficiency of EfOM is dependent on the composition of secondary effluent, and coagulation treatment of the actual wastewater with complex organic matters needs further confirmation.

Al-based salts have also been widely applied as coagulants and flocculating agents to reduce the EfOM from secondary effluents due to their simplicity, cost-efficiency, and upgradeability. Moreover, Al-based coagulants have good natural organic matter (i.e., humic substances) removal capacity [10], with great potential for EfOM removal, which can produce higher quality water for reuse. Previous studies have reported the characteristics and mechanisms of coagulation using Al-based coagulants. For example, during humic acid removal, it is first adsorbed on the surface of $Al(OH)_{3(s)}$ formed from aluminum hydrolysis, followed by surface complexation between humic acid molecules and surface groups on $Al(OH)_{3(s)}$, under neutral condition [11]. However, secondary effluents have several complex organics. Therefore, it is important to understand the changes in SMPs during coagulation and the changes in the characteristics of humic-like substances with different coagulant dosages. This research should be performed with the actual effluent in an industrial-scale treatment plant for successful practical application.

In this study, an industrial-scale food-processing wastewater treatment plant, which used alum-based coagulation (i.e., $Al_2(SO_4)_3$) and flocculant (i.e., anionic polyacrylamide; APAM) for the enhancement of secondary effluent quality to meet the local standards for water reuse was investigated to understand the EfOM removal feasibility and cost economy of the coagulation process for treating secondary effluent. The EfOM removal efficiency and change in relative parameters (e.g., turbidity and UV_{254}) after coagulation were evaluated. The variation in organic components was monitored based on three-dimensional excitation-emission-matrix (3D-EEM) fluorescence spectra and parallel factor (PARAFAC) analysis after adding different dosages of coagulant. The influence of coagulant dosage on humic-like and SMP-like substances was detected by fluorescence quenching combined with the Ryan–Weber model and the two-dimensional correlation (2D-COS) spectroscopy calculated from synchronous fluorescence (SF) spectra. Structural changes in secondary proteins in the EfOM before and after coagulation were analyzed using Fourier transform infrared (FT-IR) spectroscopy. The effect of $Al_2(SO_4)_3$ and PAM on the effluent quality after coagulation/flocculation treatment was assessed. Finally, the cost of the treatment was evaluated, and the actual performance of secondary effluent treatment was compared, which provided a basis for promoting the utilization of wastewater resources and improvement of the efficiency and quality of water reuse.

2. Material and Methods

2.1. Experiment Setup and Operation

In this study, a wastewater treatment plant of a food-processing factory located in Beijing, China, with a stable operation for more than five years, was studied. The mainstream biological units comprise a microaerobic reactor coupled with a membrane bioreactor (MBR). The secondary effluent (Q = 20 m^3 h^{-1}; 150 m^3 d^{-1}) from the MBR then flowed into the coagulation/flocculation reaction section, which was composed of a coagulation reactor (effective volume = 7.1 m^3; hydraulic retention time (HRT) = 20 min; mixing = 135 rpm × 2.2 kw), a neutralization reactor (effective volume = 7.1 m^3; HRT = 20 min; mixing = 135 rpm × 2.2 kw), a flocculation reactor (effective volume = 7.1 m^3; HRT = 20 min; mixing = 60 rpm × 2.2 kw), and a sedimentation tank (effective volume = 155 m^3). $Al_2(SO_4)_3 \cdot 18\ H_2O$ and APAM were selected as the coagulant and flocculant with concentrations of 150 g m^{-3} and 1.5 g m^{-3}, respectively, which were dissolved completely before adding into the coagulation and flocculation reactor. The addition of NaOH was not required because the pH of secondary effluent was around 7–8 during data collection. The effluent quality met the national standard: "The reuse of urban recycling water-water quality standard for industrial uses" (GB/T 19923-2005).

2.2. Jar Test Procedure

Two jar tests were performed in this study. First, the effluent was treated with different dosages of $Al_2(SO_4)_3 \cdot 18\ H_2O$ (i.e., evenly divided into 10 groups from 0.09 to 0.94 mmol Al L^{-1}), and the change in fluorescence characteristics of EfOM was determined. Each sample was filtered through 0.45 μm glass fiber filters (Whatman, Little Chalfont, Buckinghamshire, UK) before analysis. Secondly, the combined effect of $Al_2(SO_4)_3$ and APAM on the effluent quality after coagulation/flocculation treatment was evaluated. Five different dosages of $Al_2(SO_4)_3 \cdot 18\ H_2O$ (i.e., 110, 125, 140, 155, and 170 mg L^{-1}) were selected based on the variation in the range of the actual addition during the wastewater treatment. The ratio of APAM to $Al_2(SO_4)_3 \cdot 18\ H_2O$ was 0.012, 0.011, 0.010, 0.009, and 0.008, respectively, for each dosage of coagulant, and these ratios were also determined based on the actual range of addition. All the experimental conditions were consistent with the actual plant operating conditions, except that the settling time for the experiments was 5 min. After settling, the supernatant of each treated group was collected and analyzed.

2.3. FT-IR

FT-IR spectra were measured using an FT-IR spectrophotometer (Perkin Elmer Frontier, Waltham, MA, USA) by scanning between wavenumbers of 4000–400 cm^{-1}. Each sample was frozen at −33 °C in an ultra-low temperature freezer (Haier, DW-86L338, Qingdao, China) and then dewatered with a freezer dryer at −57 °C and 5 Pa. All FT-IR samples were prepared by pressing a homogenized mixture of 1 mg of freeze-dried sample and 140 mg of potassium bromide (spectrum pure reagent).

2.4. Dissolved Organic Matter (DOM) Fluorescence Characteristics Analysis

2.4.1. Three-Dimensional Excitation-Emission-Matrix (3D-EEM) and Parallel Factor Analysis (PARAFAC)

Each sample for dissolved organic matter (DOM) analysis was collected by filtration using 0.45 μm glass fiber filters (Whatman, Little Chalfont, Buckinghamshire, UK). The

3D-EEM spectra of DOM samples were acquired using a fluorescence spectrophotometer (F-7000, Hitachi, Tokyo, Japan). The excitation spectra were scanned every 5 nm from 200 to 450 nm, and the emission spectra were scanned every 5 nm from 200 to 550 nm. Fluorescence regional integration (FRI) was used for calculating the proportion of the five types of fluorescent DOM (FDOM), measured by 3D-EEM. The fluorescence proportion of region i ($P_{i,n}$, %) was calculated according to Equation (1):

$$P_{i,n}\ (\%) = \frac{\varnothing_{i,n}}{\varnothing_{T,n}} \times 100\% = \frac{MF_i \sum_{ex} \sum_{em} I(\lambda_{ex}\lambda_{em})\Delta\lambda_{ex}\Delta\lambda_{em}}{\sum_{i=1}^{5} \varnothing_{i,n}} \times 100\%,\ i = I - V \quad (1)$$

where $\varnothing_{i,n}$ is the Ex/Em area referring to the value of region i; $\varnothing_{T,n}$ is the Ex/Em area referring to the value of the total region; MF_i is a multiplication factor for each region; I ($\lambda_{ex}\lambda_{em}$) is the fluorescence intensity for each Ex/Em wavelength pair; $\Delta\lambda_{ex}$ and $\Delta\lambda_{em}$ are Ex and Em intervals, respectively [12].

PARAFAC was calculated using the DOMFluor toolbox (version 1.7) and N-way toolbox in MATLAB software (MATLAB2021, Natick, MA, USA). The data collected from 44 experimental samples were imported into MATLAB. Modeling was achieved by a series of step-wise procedures performed in a previous study [13], several FDOM components were identified, and the maximum fluorescence intensity (Fmax) was used to characterize the relative concentrations of individual FDOM components [14].

2.4.2. Fluorescence Quenching Titration

The Ryan–Weber model, a commonly used nonlinear fitting model, was used to calculate the coagulation complex constant of FDOM samples and Al-coagulant. This model is mainly based on the following assumptions: (1) the fluorescence quenching of each FDOM component is assumed to be the result of complexation with Al, (2) Al and FDOM have the same and independent binding site or ligand binding, and (3) the stoichiometric ratio of Al to ligand complex is 1:1. The calculation of the R-W model is based on Equation (2):

$$I = I_0 + (I_{ML} - I_0)(\frac{1}{2K_M C_L})(1 + K_M C_L + K_M C_M - \sqrt{(1 + K_M C_L + K_M C_M)^2 - 4{K_M}^2 C_L C_M}) \quad (2)$$

$$f = \frac{(I_0 - I_{ML})}{I_0} \quad (3)$$

where I_0 is the DOM fluorescence intensity of the water sample without Alum; C_M represents the Alum concentration; I represents the fluorescence intensity when the Alum concentration is C_M; C_L represents the complexing capacity concentration of Al (mmol L^{-1}), and K_M is the Al complexing stability constant. At the same time, I_{ML} indicates the concentration threshold of Al when the fluorescence intensity stabilizes; f is expressed as the quenching rate, that is, the ratio of the binding group to the total group that was calculated using Equation (3) [15].

2.4.3. Two-Dimensional Synchronous Fluorescence Correlation Spectroscopy (2D-SF-COS)

The synchronous fluorescence (SF) spectra were recorded in the range of 250–550 nm with a constant offset ($\Delta\lambda$ = 60 nm) using a fluorescence spectrophotometer (F-7000, Hitachi, Tokyo, Japan) with a scanning speed of 240 nm min^{-1} and a 1 nm scan spacing [15]. The SF spectra with different Al-based coagulant concentrations (see Section 2.2) were used to perform the two-dimensional correlation spectroscopy (2D-COS) using "2D-Shige" software (Kwansei-Gakuin University, Japan).

2.5. Analytical Methods

Samples collected from the influent and effluent of the coagulation/flocculation-settling process were analyzed for the different COD fractions, i.e., total COD (TCOD), particulate COD (pCOD; >1.5 μm), colloidal COD (cCOD; 0.45–1.5 μm), and soluble COD (sCOD; <0.45 μm), for which the samples were filtered through 0.45 and 1.5 μm glass-fiber filters [16]. All COD concentrations were measured following the Chinese standard method [17]. The suspended solids (SS) concentration was determined after filtration with 0.45 μm glass-fiber filters. The pH was measured using a portable multimeter (PHB-4, Zsynet, Shanghai, China). The concentration of dissolved organic carbon was quantified using a total organic carbon analyzer (MULTI N/C® 3100, Analytik Jena, Thuringia, Germany). Turbidity was measured using a turbidimeter. UV_{254} was determined by UV absorbance at 254 nm wavelength with a UV–visible–NIR spectrophotometer (LH-3BA, Shenzhen, China). Specific ultraviolet absorbance (SUVA) was calculated using Equation (4) [18]:

$$SUVA\left(\mathrm{L\,mg^{-1}m^{-1}}\right) = \frac{UV_{254\mathrm{nm}}}{DOC} \times 100 \quad (4)$$

Chroma is measured as color number (CN) with the absorbance measured at 436 nm (A_{436}), 525 nm (A_{525}), and 620 nm (A_{620}) wavelengths, and calculated using Equation (5) [19]:

$$CN\left(\mathrm{cm^{-1}}\right) = \frac{A_{436}^2 + A_{525}^2 + A_{620}^2}{A_{436} + A_{525} + A_{620}} \times 100 \quad (5)$$

3. Results and Discussion

3.1. Performance of the Coagulation Process

The changes in organic matter and wastewater characteristics before and after coagulation are shown in Table 1. The TCOD concentration in the influent was 26.36 mg L^{-1}, and that in the effluent was 14.14 mg L^{-1}. The average removal efficiency of TCOD was 44.61%. The UV_{254} and SUVA in the influent were 0.263 and 2.25 cm^{-1}, which decreased by 25.13% and 9.13% to 0.197 and 2.07 cm^{-1}, respectively, in the effluent after coagulation/flocculation treatment.

Table 1. The performance of coagulation/flocculation process under treating secondary effluent.

Parameter	Influent	Effluent	Removal Efficiency (%)
TCOD (mg L^{-1})	26.36 ± 7.20	14.41 ± 3.57	44.61 ± 6.53
UVA_{254} (cm^{-1})	0.263 ± 0.002	0.197 ± 0.005	25.13 ± 1.64
SUVA (L mg^{-1} m^{-1})	2.25 ± 0.03	2.07 ± 0.03	9.13 ± 1.35
CN (cm^{-1})	0.020 ± 0.001	0.012 ± 0.002	36.65 ± 9.49
SS (mg L^{-1})	4.56 ± 1.50	0.66 ± 0.58	85.01 ± 12.00
Turbidity (NTU)	1.19 ± 0.28	0.63 ± 0.45	50.35 ± 34.43
pH	7.92 ± 0.28	7.00 ± 0.05	11.52 ± 1.25

TCOD, total COD; UVA_{254}, ultraviolet absorbance at 254 nm; SUVA, specific ultraviolet absorbance; CN, color number; SS, suspended solids.

Compounds that absorb at 254 nm comprise aromatic groups that are often considered biorecalcitrant [20], and the decrease in UV_{254} value after coagulation treatment indicates significantly improved effluent quality. The percentage reduction in UV_{254} absorbing components was lower than that of TCOD, which indicates a more efficient removal of nonaromatic organic matter compared to aromatic constituents [20] after treatment. The lower SUVA value of effluent compared to that of influent indicated its lower aromaticity and the possibility of achieving a lower high-molecule-weight refractory humic-like compounds content through coagulation [8].

The values for the other parameters (i.e., pH, SS, CN, and turbidity) in the influent and effluent are depicted in Table 1. The pH decreased from 7.92 to 7.00, indicating that the process of coagulation/flocculation reduced the pH value of the alkaline influent to neutral. The SS concentration in the influent was 4.56 mg L^{-1}, which decreased by 85.01% to 0.66 mg L^{-1} in the effluent after coagulation/flocculation. Because of the efficient removal of SS and TCOD, the CN and turbidity were significantly reduced by 36.65% and 50.35%, respectively, after the coagulation/flocculation process. The turbidity removal efficiency was better than that of TCOD, which was attributed to the difference in coagulation mechanisms of particulate, colloidal, and dissolved organic matter [7]. This suggested that organics contributing to the turbidity were removed efficiently. These phenomena were attributed to the efficient removal of organic matter from secondary effluent.

3.2. Change in COD Fractions

Three types of COD fractions (i.e., pCOD, cCOD, and sCOD) were monitored, as shown in Figure 1.

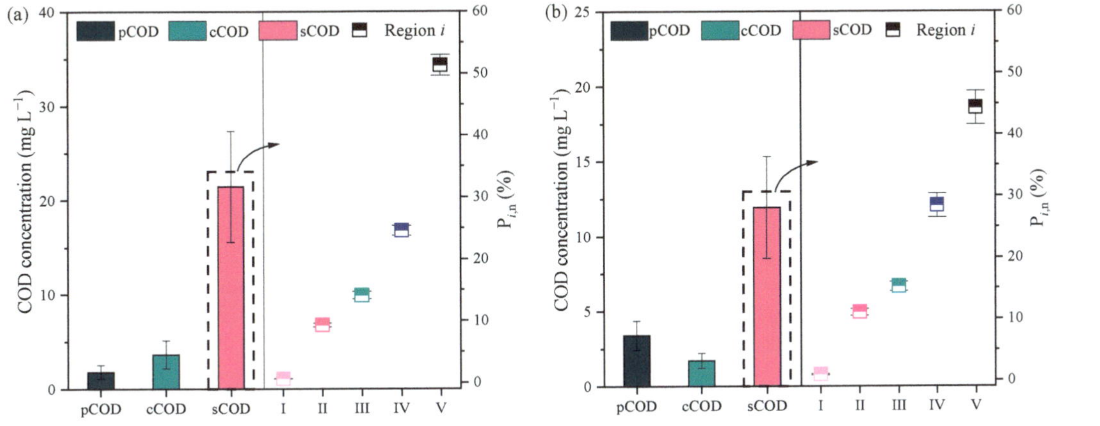

Figure 1. The concentration of COD fractions and organic composition of sCOD (solubleCOD) in the influent (**a**) and effluent (**b**).

The concentration of pCOD and cCOD in the influent was 1.81 mg L^{-1} and 3.63 mg L^{-1}, and those in the effluent were 3.41 mg L^{-1} and 1.71 mg L^{-1}, respectively. The contribution of these two types of COD was very less in the TCOD, with an average of 23.06% and 30.03% before and after coagulation. The sCOD decreased from 21.47 mg L^{-1} to 11.93 mg L^{-1}, which significantly contributes to TCOD removal. The pH of the influent was approximately 7.9 in this study. According to a previous study report [21], prehydrolyzed aluminum salt solutions may react with the alkalinity present in EfOM, as shown below:

$$Al_2(SO_4)_3 + 6OH^- \rightarrow 2Al(OH)_3 + 3SO_4{}^{2-}$$

At this pH, the Al hydroxide precipitates, completely dominating all other Al species. Thus, the major coagulation mechanisms for the removal of pCOD and cCOD are adsorption on the precipitates and/or sweep flocculation [22]. The concentration of pCOD in the effluent increased slightly due to the small flocs with adsorbed organic matter remaining suspended in the effluent. The sCOD (i.e., DOM), the major fraction of EfOM, decreased significantly by 44.43%, and changes in its characteristics during treatment were further examined.

In order to understand the changes in DOM components, FRI analysis based on EEM fluorescence spectral data was performed. Five types of DOM components were identified by FRI analysis. Region I and II contents were identified as protein-like components, constituting 0.67% and 9.31% of DOM in the influent, respectively, showing a slight increase after treatment. Region III, a fulvic-like component, changed slightly from 14.08% to 15.23% after coagulation. Region IV and V constituted the highest proportions in the influent and effluent, changed notably with the addition of alum coagulant. Region IV (i.e., SMP-like component) increased from 24.61% to 28.44%, while Region V humic-like substance decreased significantly from 51.32% to 44.35%. These results confirmed that humic-like substances could be efficiently removed by alum [23], and it could contribute to maintaining a low CN in the effluent. Alum-based coagulants have superior natural organic matter (NOM; e.g., humic-like substance) removal capacity based on sweep adsorption and surface complexation [10,11]. However, alum did not seem to have a significant effect on the removal of protein-like components, including SMP. It may be attributed to the fact that SMP-like compounds were more hydrophilic and hence more difficult to remove by Al-based coagulants [24]. To some extent, the significant reduction in humic-like substances caused an increase in the proportions of other substances.

3.3. *Influence on DOM Components*

3.3.1. Component Separation and Dehumification Analysis

Three principal components were identified from the 3D-EEM fluorescence spectra using PARAFAC analysis, and the contour plots of the three components are shown in Figure 2a–c.

The fluorescence component 1 (FC1; Figure 2a), with two peaks at wavelengths λex/λem = 315 (245) nm/400 nm, was assigned to microbial humic-like fluorescence [8]. The fluorescence component 2 (FC2; Figure 2b) with a peak at λex/λem = 365 (260) nm/450 nm was identified as a large-molecular-size humic-like compound [8]. The fluorescence component 3 (FC3; Figure 2c) was a protein-like substance associated with microbial activity and biological productivity that could be regarded as SMP, with a fluorescence peak at 280 nm/250 nm (λex/λem). The analysis indicated that there was a considerable amount of allochthonous organic matter (i.e., humic-like material) in EfOM, and the SMPs were less than the humic-like materials [8]. It was also suggested that these two substances significantly influence EfOM, which was consistent with a report from a previous study [25].

In order to further evaluate the removal efficiency of each component and the effectiveness of dehumification during the addition of alum-based coagulation, fluorescence quenching titration and synchronous fluorescence spectra were established (Figure 2d). There were considerable differences in the changes in DOM components with changes in concentration of $Al_2(SO_4)_3$. The Fmax of the two humic-like components (i.e., FC1 and FC2) decreased gradually with increasing Al addition. The Fmax of FC1 was much higher than that of FC2, which indicated that microbial humic-like substances are major constituents of EfOM in this study. Unlike the variations in FC1 and FC2, the change in Fmax of FC3 was inconsistent at different dosages of coagulant, which meant the removal efficiency of SMP could not be accurately estimated by fluorescence intensity using EEM spectra. The value of A3 (400–550 nm)/A1 (300–350 nm) was positively correlated with humification degree [26], and it decreased from 0.17 to 0.12 when Al concentration increased from 0.00 to 0.94 mmol L^{-1}. This indicated that the humific acid content of effluent gradually decreased when FC1 and FC2 were efficiently removed from EfOM.

The two types of hydrophobic organic components (i.e., FC1 and FC2) can interact with $Al_2(SO_4)_3$ through complexation with hydroxyl ions on the surface [7,8]. The Log Km and f values of FC1 and FC2 are shown in Table 2; according to the Ryan–Weber model fitting, the different removal characteristics between FC1 and FC2 were examined. The different contributions of the two types of humic-like substances on the dehumification of the secondary effluents during the process of coagulation were evaluated. For FC1, the values of Log Km and f were 4.12 and 0.12, respectively; those for FC2 were 4.10 and 0.22. It was suggested that FC1 was more sensitive to Al-based coagulant, compared to FC2, either from the aspect of complex constants or from the degree of quenching. Hence, it can be confirmed that microbial humic-like substances can be efficiently removed by $Al_2(SO_4)_3$

addition, and large-molecular-size humic-like compounds were also complex during this process, although their removal efficiency was not as effective as that of the former.

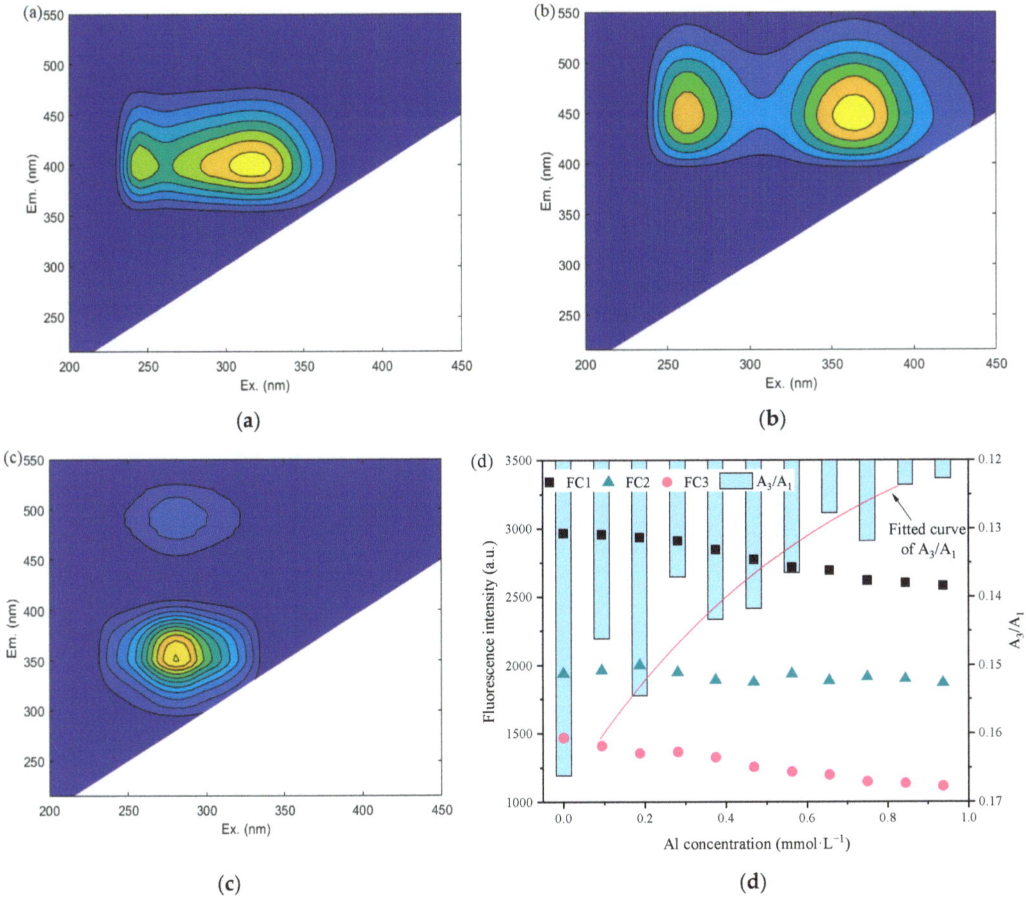

Figure 2. The EEM−PARAFAC characteristics of FC1 (**a**), FC2 (**b**), and FC3 (**c**) of DOM, and (**d**) Fmax and humification degree changes after treatment with different Al concentrations.

Table 2. Complexing stability constant of Al with DOM by the Ryan–Weber model.

Component	Log Km	f	R^2
FC1	4.12	0.12	0.9197
FC2	4.10	0.22	0.7289

3.3.2. Variation in SMP

The effect of $Al_2(SO_4)_3$ on the forms and contents of the corresponding functional groups (i.e., C=O and N-H) of SMP-like substances, and the change in the composition ratio of the related secondary structures, were examined by FTIR analysis. FTIR spectra at wavelength range 1700–1600 cm^{-1} (i.e., amide I region) are shown in Figure 3.

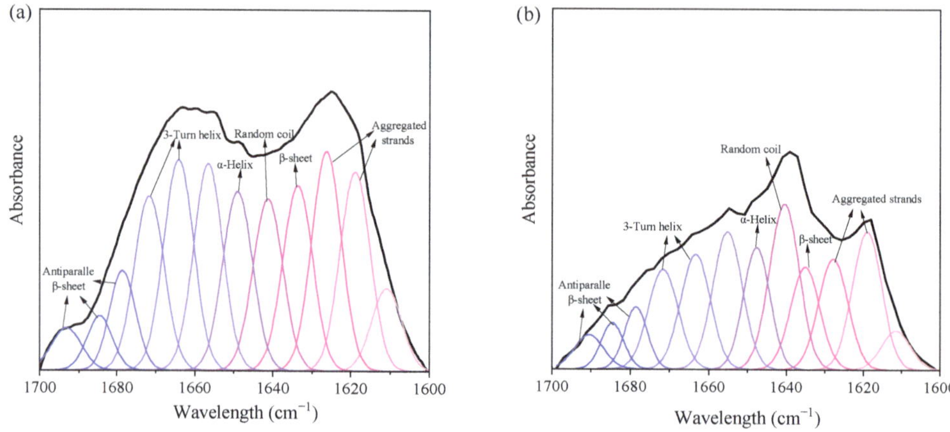

(**a**) (**b**)

Figure 3. FTIR spectra and Second derivative resolution enhanced and curve−fitted amide I region (1700−1600 cm^{-1}) for proteins of the samples: (**a**) influent; (**b**) effluent.

Several types, i.e., aggregated strands, β-sheet, random coil, α-helix, 3-turn helix, and antiparallel β-sheet/aggregated strands, were sorted from secondary structures of SMP protein [27]. Results confirmed that adding $Al_2(SO_4)_3$ reduced the total areas of peaks, and the cutback in the total areas of peaks indicated an effective reduction in protein content (Table 3). The proportions of α-helix and random coil increased from 10.39% and 10.43% to 12.24% and 15.77%, respectively, while that of β-Sheet decreased from 10.43% to 8.95%. The ratio of α-helix to (β-sheet + random coil) was 0.498 in the influent, and in the effluent, it was 0.495. A high percentage of α-helix and a low percentage of β-sheet could lead to the tight structure of the protein, while the low value of the ratio of α-helix and (β-sheet + random coil) confirmed that the loose structure of protein molecule promoted more hydrophobicity in SMP protein [27]. Hence, coagulation with $Al_2(SO_4)_3$ could reduce the proteins in SMP from EfOM, aided by a loose SMP protein structure and enhanced hydrophobicity of SMP proteins.

Table 3. Band assignments for the protein secondary structures of dissolved organic matter.

Wavelength cm^{-1}	Secondary Structures	Influent	Effluent
		Proportion of Area (%)	Proportion of Area (%)
1616–1626	Aggregated strands	20.86	18.04
1629–1640	β-sheet	10.43	8.95
1640–1645	Random coil	10.43	15.77
1649–1650	α-Helix	10.39	12.24
1655–1669	3-Turn helix	20.78	17.20
1676–1694	Antiparallel β-sheet/aggregated strands	14.21	12.80
	α-Helix/(β-sheet +random coil)		

3.3.3. Characteristics of DOM Removal

To understand the changes in DOM with different dosages of $Al_2(SO_4)_3$ during coagulation, synchronous and asynchronous maps were generated from SF spectra. Three major auto-peaks were revealed in the synchronous map (Figure 4a). The change in band intensity followed this order: 330 > 362 > 280 nm. The bands at 330 nm and 362 nm were related to humic-like fluorescence compounds, and the former band may be derived from microbial processes as microbial humic-like components [15,28]. This analysis showed that the DOM changed with time in the following order: humic-like > protein-like substances and microbial humic-like components have an advantage over other humic-like substances in the process of coagulation. This result was consistent with the EEM analysis of dehumification that the removal of microbial humic-like substance plays a major role in this process. Furthermore, the cross-peak of the two bands showed a positive correlation, implying that the three component signals consistently changed during the process of coagulation with $Al_2(SO_4)_3$.

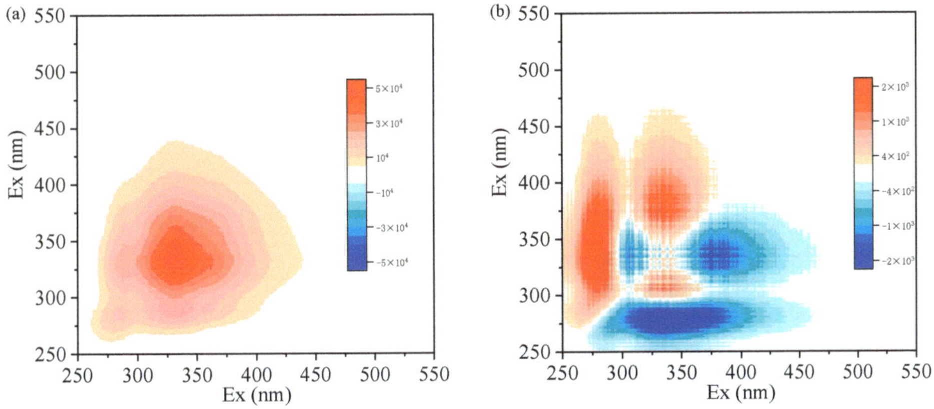

(a) (b)

Figure 4. Two−dimensional correlation maps generated based on SFS of DOM samples treated with different Al concentrations: (**a**) 250−550 cm^{-1} synchronous spectra; (**b**) 250−550 cm^{-1} asynchronous spectra. Red represents positive correlations, and blue represents negative correlations; higher color intensity indicates a stronger positive or negative correlation.

The asynchronous map displays distinctive characteristics compared with the synchronous map, which reveals the degree of the sequential changes at the three different wavelengths with the addition of $Al_2(SO_4)_3$ (i.e., the extent to which the fluorescence intensity changes at one wavelength either leads or lags behind that at the other wavelength) [28], shown in Figure 4b. According to Noda's rule [29], the changes in the bands with time followed this order: 330 > 362 > 280 nm. These results imply that microbial humic-like fractions might be removed earlier than the other fractions, and the protein-like substances displayed a weaker capturing capacity by Al-based coagulant (i.e., $Al_2(SO_4)_3$). Hence, humic-like components of EfOM could have more chances of getting removed by $Al_2(SO_4)_3$ than other fractions during coagulation.

3.4. Optimization of Coagulation/Flocculation Efficiency

The concentration of $Al_2(SO_4)_3 \cdot 18\ H_2O$ and APAM influence the efficiency of EfOM removal during coagulation and flocculation processes. The change in EEM-PARAFAC characteristics of FC1, FC2, and FC3 of FDOM in EfOM treated by different concentrations of $Al_2(SO_4)_3 \cdot 18\ H_2O$ and APAM is exhibited in Figure 5a. The three components had blue shifts in different degrees due to the gradually increased concentration of $Al_2(SO_4)_3 \cdot 18\ H_2O$ and APAM, which positively influenced FDOM removal. With the change in concentration of these two treatment chemicals, based on the range of concentration applied in this wastewater treatment plant for EfOM removal, the three components of FDOM exhibited different removal characteristics. When the concentration of coagulant was around 110–120 mg L^{-1}, the Fmax of FC1, FC2, and FC3 decreased notably with the increase in the ratio of flocculant to coagulant concentration. The flocculant concentration would influence the efficiency of microbial humic-like substance removal in this situation.

However, the influence of a higher proportion of flocculant on three components removal was not always positive. Its efficiency stabilized when the dosage of flocculant exceeded a certain amount (i.e., the ratio of APAM to $Al_2(SO_4)_3 \cdot 18\ H_2O$ was 0.009 in this study). When the concentration of coagulant increased, the Fmax of humic-like substances decreased significantly compared to protein-like substances; humic-like components located at the high band of Ex/Em, in particular, displayed a reduction rate of Fmax up to 14.10%, in response to increased concentrations of coagulant combined with an equal proportion of the flocculant. However, it has been reported that a higher concentration of PAM might form large amounts of loose-structure flocs that negatively affect organics removal [30]. Hence, the removal efficiencies of the three components did not continue to increase with the increase in the concentration of $Al_2(SO_4)_3 \cdot 18\ H_2O$; and did not change significantly when the concentration of $Al_2(SO_4)_3 \cdot 18\ H_2O$ exceeded 155 mg L^{-1}.

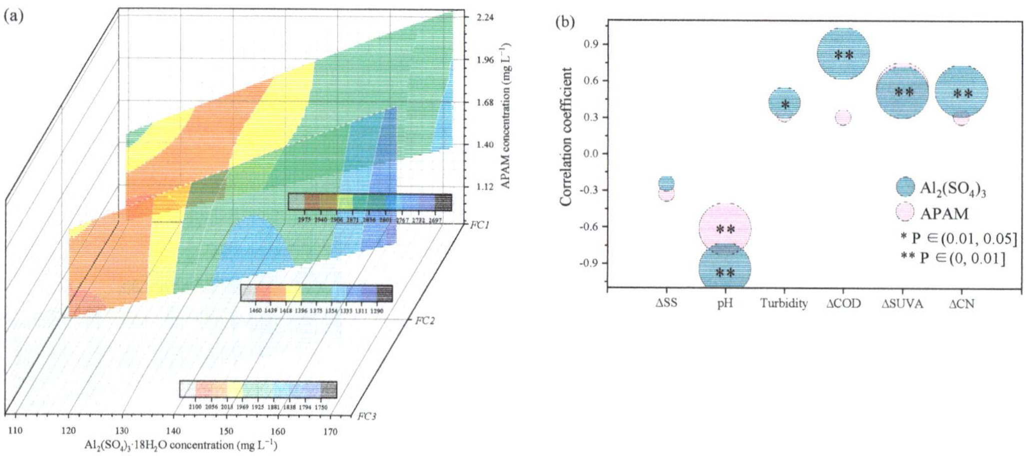

(**a**) (**b**)

Figure 5. The influence of coagulant and flocculant concentration on FC1, FC2, and FC3 components of DOM (**a**) and effluent pollutant indices (**b**).

The correlation between the concentration of $Al_2(SO_4)_3 \cdot 18\ H_2O$ and APAM and the efficiency of coagulation is shown in Figure 5b to demonstrate the effect of coagulant and flocculant treatment on pollutant removal. The removal rate of SS was not proportional to the concentration of coagulant and flocculant, and it may be attributed to the effects of MBR treatment which results in very less SS in secondary effluent. The turbidity was significantly negatively correlated ($p < 0.05$) with the concentration of coagulant while the removal rate of COD, SUVA, and CN was significantly positively correlated ($p < 0.01$) with the coagulant concentration. It indicated coagulation by $Al_2(SO_4)_3$ resulted in a major reduction in EfOM from a secondary effluent and efficiently guaranteed the effluent quality of WWTP. APAM, as a flocculant, balances the pH and reduces SUVA contents. These results also suggested that the concentrations of coagulant and flocculant were positively related to the effluent quality in the range of dosage applied in the study, and the adjustment of chemical dosages can effectively address the changes in secondary effluent quality (i.e., variation in EfOM).

3.5. Integrated System Economic Evaluation

The cost of building or upgrading treatment systems is one of the most critical barriers to the widespread utilization of industrial wastewater reuse [31]. The total cost for applying a coagulation/flocculation and sedimentation integrated treatment system for the secondary effluent treatment of the industrial-scale food-processing wastewater treatment plant was estimated. Table 4 summarizes the resources used and the cost calculation.

Table 4. Total capital investment for coagulation process of secondary effluent treatment.

Parameter	Value
Q	150 m^3 d^{-1}
No. of working days per year	365 d
tw	7 h d^{-1}
Construction cost (designing & treatment units manufacturing & mechanical cost)	CNY 4,000,000
Electrical costs	CNY 0.98 t^{-1}
Chemical costs $Al_2(SO_4)_3 \cdot 18\ H_2O$	CNY 0.45 t^{-1}
APAM	CNY 0.08 t^{-1}
Maintenance cost	CNY 5000 a^{-1}

The construction cost (manufacturing of treatment tanks and mechanical cost) was estimated to be 4,000,000 CNY, and the electrical power cost was estimated at 0.98 CNY t^{-1}. On the other hand, the chemical costs were estimated to be about 0.45 CNY t^{-1} for the coagulant and 0.08 CNY t^{-1} for the flocculant. The maintenance cost was considered to be 5000 CNY a^{-1}. The average cost of secondary effluent treatment from this study was 1.51 CNY t^{-1} with a 44.61% COD removal rate, which implies a cost of 0.034 CNY t^{-1} $\%COD^{-1}$; Su et al. [2] had applied Fenton coupled with biological aerated filter to treat EfOM, and this treatment cost was 0.045 CNY t^{-1} $\%COD^{-1}$ to meet the effluent standard, which was higher than that estimated in this study. Moreover, it is also more economical compared with ozonation (estimated cost: 0.03–0.07 CNY t^{-1} $gCOD^{-1}$) to improve effluent quality [32]. Compared to other processes [33], Al-based coagulant has the advantages of low cost and high efficiency and has great practical application potential in water reuse for industrial uses. Overall, $Al_2(SO_4)_3$ and APAM-based coagulation/flocculation is a comparatively economic treatment process for efficient EfOM removal to realize food-processing wastewater reuse.

4. Conclusions

In this study, $Al_2(SO_4)_3$ and APAM were chosen as coagulant and flocculant, respectively, for the treatment of secondary effluent from an industrial-scale food-processing wastewater treatment plant to meet the regulatory standards for wastewater reuse. The EfOM was removed efficiently with a 44.61% COD removal rate during this process, and both dearomatization and dehumification were achieved to different degrees. Coagulation played a key role in the removal of humic-like components of EfOM, especially the organics from microbial humic-like components. The degree of removal of protein-like components of SMP was lower than humic-like components; however, coagulation can produce a loose SMP protein structure and enhance the hydrophobicity of SMP proteins. Flocculation can further reduce the aromaticity of the secondary effluent and strengthen the sedimentation. The cost of secondary effluent treatment in this study was 0.034 CNY t^{-1} $\%COD^{-1}$, which showed that Al-based coagulant had a viable application value for EfOM removal to reach the goal of food-processing wastewater reuse.

Author Contributions: Y.C.: Methodology, Experiment, Writing—Original Draft, Conceptualization; Q.C.: Experiment; C.Z.: Experiment; X.R.: Writing—Review & Editing, Supervision, Validation; Y.W.: Supervision; Y.K.: Supervision; K.C.: Review; M.-H.K.: Review; M.-H.H.: Writing—Reviewing and Editing, Supervision, Validation. All authors have read and agreed to the published version of the manuscript.

Funding: This work was supported by the Korea Agency for Infrastructure Technology Advancement (KAIA) grant funded by the Ministry of Land, Infrastructure and Transport (Grant RS-2022-00144137), and Beijing Natural Science Foundation (8222043).

Institutional Review Board Statement: Not applicable.

Informed Consent Statement: Informed consent was obtained from all subjects involved in the study.

Data Availability Statement: The data presented in this study are available on request from the corresponding author.

Conflicts of Interest: The authors declare no conflict of interest. The paper reflects the views of the scientists, and not the company.

References

1. Wang, Y.; Jin, X.; Zhuo, N.; Zhu, G.; Cai, Z. Interaction-sedimentation strategy for highly efficient removal of refractory humic substances in biologically treated wastewater effluent: From mechanistic investigation to full-scale application. *J. Hazard. Mater.* **2021**, *418*, 126145. [CrossRef] [PubMed]
2. Su, T.; Wang, Z.; Zhou, K.; Chen, X.; Cheng, Y.; Zhang, G.; Wu, D.W.; Sun, S.P. Advanced treatment of secondary effluent organic matters (EfOM) from an industrial park wastewater treatment plant by Fenton oxidation combining with biological aerated filter. *Sci. Total Environ.* **2021**, *784*, 147204. [CrossRef] [PubMed]
3. Liu, J.W.; Zhao, M.F.; Duan, C.; Yue, P.; Li, T.G. Removal characteristics of dissolved organic matter and membrane fouling in ultrafiltration and reverse osmosis membrane combined processes treating the secondary effluent of wastewater treatment plant. *Water Sci. Technol.* **2021**, *83*, 689–700. [CrossRef] [PubMed]
4. Zhu, X.F.; Liu, J.D.; Li, L.; Zhen, G.Y.; Lu, X.Q.; Zhang, J.; Liu, H.B.; Zhou, Z.; Wu, Z.C.; Zhang, X.D. Prospects for humic acids treatment and recovery in wastewater: A review. *Chemosphere* **2023**, *312*, 137193. [CrossRef]
5. Qu, J.H.; Dai, X.H.; Hu, H.Y.; Huang, X.; Chen, Z.; Li, T.; Cao, Y.S.; Daigger, G.T. Emerging Trends and Prospects for Municipal Wastewater Management in China. *ACS EST Eng.* **2022**, *2*, 323–336. [CrossRef]
6. De Oliveira Anício, S.; dos Santos Lopes, V.; de Oliveira, A.L. PSD and Fractal Dimension for flocculation with different parameters and ferric chloride, aluminium polychloride and aluminium sulfate as coagulants. *J. Water Process Eng.* **2021**, *43*, 102180. [CrossRef]
7. Xiong, X.; Wu, X.; Zhang, B.; Xu, H.; Wang, D. The interaction between effluent organic matter fractions and $Al_2(SO_4)_3$ identified by fluorescence parallel factor analysis and FT-IR spectroscopy. *Colloids Surf. A Physicochem. Eng. Asp.* **2018**, *555*, 418–428. [CrossRef]
8. Yu, H.; Qu, F.; Sun, L.; Liang, H.; Han, Z.; Chang, H.; Shao, S.; Li, G. Relationship between soluble microbial products (SMP) and effluent organic matter (EfOM): Characterized by fluorescence excitation emission matrix coupled with parallel factor analysis. *Chemosphere* **2015**, *121*, 101–109. [CrossRef]
9. Liu, Y.Y.; Zhang, W.J.; Yang, X.Y.; Xiao, P.; Wang, D.S.; Song, Y. Advanced treatment of effluent from municipal WWTP with different metal salt coagulants: Contaminants treatability and floc properties. *Sep. Purif. Technol.* **2013**, *120*, 123–128. [CrossRef]
10. Xu, Y.; Chen, T.; Cui, F.; Shi, W. Effect of reused alum-humic-flocs on coagulation performance and floc characteristics formed by aluminum salt coagulants in humic-acid water. *Chem. Eng. J.* **2016**, *287*, 225–232. [CrossRef]
11. Jin, P.; Song, J.; Wang, X.C.; Jin, X. Two-dimensional correlation spectroscopic analysis on the interaction between humic acids and aluminum coagulant. *J. Environ. Sci.* **2018**, *64*, 181–189. [CrossRef]
12. Song, F.; Wu, F.; Feng, W.; Tang, Z.; Giesy, J.P.; Guo, F.; Shi, D.; Liu, X.; Qin, N.; Xing, B.; et al. Fluorescence regional integration and differential fluorescence spectroscopy for analysis of structural characteristics and proton binding properties of fulvic acid sub-fractions. *J. Environ. Sci.* **2018**, *74*, 116–125. [CrossRef] [PubMed]
13. Stedmon, C.A.; Bro, R. Characterizing dissolved organic matter fluorescence with parallel factor analysis: A tutorial. *Limnol. Oceanogr. Methods* **2008**, *6*, 572–579. [CrossRef]
14. Cheng, Y.; Tian, K.; Xie, P.; Ren, X.H.; Li, Y.; Kou, Y.; Chon, K.; Hwang, M.H.; Ko, M.H. Insights into the minimization of excess sludge production in micro-aerobic reactors coupled with a membrane bioreactor: Characteristics of extracellular polymeric substances. *Chemosphere* **2022**, *292*, 133434. [CrossRef] [PubMed]
15. Yuan, D.; Wang, H.; An, Y.; Guo, X.; He, L. Insight into the binding properties of carbamazepine onto dissolved organic matter using spectroscopic techniques during grassy swale treatment. *Ecotoxicol. Environ. Saf.* **2019**, *173*, 444–451. [CrossRef]
16. Rahman, A.; De Clippeleir, H.; Thomas, W.; Jimenez, J.A.; Wett, B.; Al-Omari, A.; Murthy, S.; Riffat, R.; Bott, C. A-stage and high-rate contact-stabilization performance comparison for carbon and nutrient redirection from high-strength municipal wastewater. *Chem. Eng. J.* **2019**, *357*, 737–749. [CrossRef]
17. *HJ828*; Water Quality-Determination of the Chemical Oxygen Demand-Dichromate Method. Ministry of Ecology and Environmental of the People's Republic of China: Beijing, China, 2017.
18. Hussain, S.; van Leeuwen, J.; Chow, C.; Beecham, S.; Kamruzzaman, M.; Wang, D.; Drikas, M.; Aryal, R. Removal of organic contaminants from river and reservoir waters by three different aluminum-based metal salts: Coagulation adsorption and kinetics studies. *Chem. Eng. J.* **2013**, *225*, 394–405. [CrossRef]
19. Hu, Y.; Gu, Z.; He, J.; Li, Q. Novel strategy for controlling colloidal instability during the flocculation pretreatment of landfill leachate. *Chemosphere* **2022**, *287*, 132051. [CrossRef]
20. Van Gijn, K.; Chen, Y.L.; van Oudheusden, B.; Gong, S.; de Wilt, H.A.; Rijnaarts, H.H.M.; Langenhoff, A.A.M. Optimizing biological effluent organic matter removal for subsequent micropollutant removal. *J. Environ. Chem. Eng.* **2021**, *9*, 106247. [CrossRef]
21. Lal, K.; Garg, A. Effectiveness of synthesized aluminum and iron based inorganic polymer coagulants for pulping wastewater treatment. *J. Environ. Chem. Eng.* **2019**, *7*, 103204. [CrossRef]

22. Prokopova, M.; Novotna, K.; Pivokonska, L.; Cermakova, L.; Cajthaml, T.; Pivokonsky, M. Coagulation of polyvinyl chloride microplastics by ferric and aluminium sulphate: Optimisation of reaction conditions and removal mechanisms. *J. Environ. Chem. Eng.* **2021**, *9*, 106465. [CrossRef]
23. Lapointe, M.; Papineau, I.; Peldszus, S.; Peleato, N.; Barbeau, B. Identifying the best coagulant for simultaneous water treatment objectives: Interactions of mononuclear and polynuclear aluminum species with different natural organic matter fractions. *J. Water Process Eng.* **2021**, *40*, 101829. [CrossRef]
24. Wang, P.; Ding, S.; An, G.; Qu, R.; Liu, X.; Fang, C.; Chu, W. Removal of disinfection by-product precursors by Al-based coagulants: A comparative study on coagulation performance. *J. Hazard. Mater.* **2021**, *420*, 126558. [CrossRef] [PubMed]
25. Tang, G.; Zheng, X.; Li, X.; Liu, T.; Wang, Y.; Ma, Y.; Ji, Y.; Qiu, X.; Wan, Y.; Pan, B. Variation of effluent organic matter (EfOM) during anaerobic/anoxic/oxic (A_2O) wastewater treatment processes. *Water Res.* **2020**, *178*, 115830. [CrossRef] [PubMed]
26. Yu, H.; Song, Y.; Xi, B.; Xia, X.; He, X.; Tu, X. Application of chemometrics to spectroscopic data for indicating humification degree and assessing salinization processes of soils. *J. Soils Sediments* **2012**, *12*, 341–353. [CrossRef]
27. Ding, A.; Lin, W.; Chen, R.; Ngo, H.H.; Zhang, R.; He, X.; Nan, J.; Li, G.; Ma, J. Improvement of sludge dewaterability by energy uncoupling combined with chemical re-flocculation: Reconstruction of floc, distribution of extracellular polymeric substances, and structure change of proteins. *Sci. Total Environ.* **2022**, *816*, 151646. [CrossRef]
28. Guo, X.-J.; Li, Y.-Z.; Feng, Y.-H.; Yuan, D.-H. Using fluorescence quenching combined with two-dimensional correlation fluorescence spectroscopy to characterise the binding-site heterogeneity of dissolved organic matter with copper and mercury in lake sediments. *Environ. Chem.* **2017**, *14*, 91–98. [CrossRef]
29. Noda, I.; Ozaki, Y. *Two-Dimensional Correlation Spectroscopy: Applications in Vibrational and Optical Spectroscopy*, 1st ed.; John Wiley & Sons: Hoboken, NJ, USA, 2004. [CrossRef]
30. Luo, Y.L.; Yang, Z.H.; Xu, Z.Y.; Zhou, L.J.; Zeng, G.M.; Huang, J.; Xiao, Y.; Wang, L.K. Effect of trace amounts of polyacrylamide (PAM) on long-term performance of activated sludge. *J. Hazard. Mater.* **2011**, *189*, 69–75. [CrossRef]
31. Meese, A.F.; Kim, D.J.; Wu, X.H.; Le, L.; Napier, C.; Hernandez, M.T. Opportunities and Challenges for Industrial Water Treatment and Reuse. *ACS EST Eng.* **2022**, *8*, 465–488. [CrossRef]
32. Hao, X.-D.; Gan, W.; Li, J.; Cao, D.-Q. Effectiveness of ozonation on oxidizing secondary effluent organic matter (EfOM) from WWTP. *China Water Wastewater* **2021**, *37*, 1–7. (In Chinese) [CrossRef]
33. Lu, H.Y.; Qian, C.Y.; Luo, S.G.; Zhu, Y.G.; Liu, R.Q.; Wu, M.R. Study on the influence and mechanism of polyferric sulfate on COD removal and reuse of scheelite flotation wastewater. *Min. Eng.* **2023**, *191*, 107940. [CrossRef]

membranes

Article

Effective Removal of Acetaldehyde Using Piperazine/Nitric Acid Co-Impregnated Bead-Type Activated Carbon

Yu-Jin Kang [1], Yu-Jin Kim [1], Seong-Jin Yoon [1], Dong-Jin Seo [1], Hye-Ryeong Cho [1], Kyeongseok Oh [2], Seong-Ho Yoon [3,*] and Joo-Il Park [1,*]

1 Department of Chemical & Biological Engineering, Hanbat National University, Daejeon 34158, Republic of Korea; yujin0581@naver.com (Y.-J.K.); yzzy0840@naver.com (Y.-J.K.); tjdwls7856@gmail.com (S.-J.Y.); tjehdwls0963@naver.com (D.-J.S.); ryeong2829@gmail.com (H.-R.C.)
2 Department of Chemical & Biological Engineering, Inha Technical College, Incheon 22212, Republic of Korea; kyeongseok.oh@inhatc.ac.kr
3 Interdisciplinary Graduate School of Engineering Science, Kyushu University, Fukuoka 816-8580, Japan
* Correspondence: yoon@cm.kyushu-u.ac.jp (S.-H.Y.); jipark94@hanbat.ac.kr (J.-I.P.); Tel.: +81-92-583-7959 (S.-H.Y.); +82-42-821-1530 (J.-I.P.)

Abstract: Acetaldehyde (CH_3CHO) in the atmosphere is associated with adverse health effects. Among the various options for use in removing CH_3CHO, adsorption is often employed because of its convenient application and economical processes, particularly when using activated carbon. In previous studies, the surface of activated carbon has been modified with amines to remove CH_3CHO from the atmosphere via adsorption. However, these materials are toxic and can have harmful effects on humans when the modified activated carbon is used in air-purifier filters. Therefore, in this study, a customized bead-type activated carbon (BAC) with surface modification options via amination was evaluated for removing CH_3CHO. Various amounts of non-toxic piperazine or piperazine/nitric acid were used in amination. Chemical and physical analyses of the surface-modified BAC samples were performed using Brunauer–Emmett–Teller measurements, elemental analyses, and Fourier transform infrared and X-ray photoelectron spectroscopy. The chemical structures on the surfaces of the modified BACs were analyzed in detail using X-ray absorption spectroscopy. The amine and carboxylic acid groups on the surfaces of the modified BACs are critical in CH_3CHO adsorption. Notably, piperazine amination decreased the pore size and volume of the modified BAC, but piperazine/nitric acid impregnation maintained the pore size and volume of the modified BAC. In terms of CH_3CHO adsorption, piperazine/nitric acid impregnation resulted in a superior performance, with greater chemical adsorption. The linkages between the amine and carboxylic acid groups may function differently in piperazine amination and piperazine/nitric acid treatment.

Keywords: bead-type activated carbon; piperazine; nitric acid; surface modification; acetaldehyde removal

Citation: Kang, Y.-J.; Kim, Y.-J.; Yoon, S.-J.; Seo, D.-J.; Cho, H.-R.; Oh, K.; Yoon, S.-H.; Park, J.-I. Effective Removal of Acetaldehyde Using Piperazine/Nitric Acid Co-Impregnated Bead-Type Activated Carbon. *Membranes* **2023**, *13*, 595. https://doi.org/10.3390/membranes13060595

Academic Editor: Daqi Cao

Received: 10 May 2023
Revised: 8 June 2023
Accepted: 10 June 2023
Published: 12 June 2023

1. Introduction

Acetaldehyde (CH_3CHO) is a volatile organic compound (VOC) in the atmosphere that is associated with adverse health effects. It is a factor in sick building syndrome, as it is found in wallpaper, furniture, cigarettes, and building materials [1–8]. Moreover, CH_3CHO is odorous at a low concentration (0.09 mg/m^3) and causes chest tightness, eye irritation, and respiratory tract irritation [9,10]. Therefore, indoor CH_3CHO content should be regulated at <0.03 ppm, and an effective removal method for CH_3CHO is required [11].

Several methods, such as absorption [12], condensation [13], biofiltration [14], and thermal oxidation [15], are commonly applied in removing VOCs such as CH_3CHO, from gas streams. These methods are effective at relatively high VOC concentrations

(>5000 ppmv) [16], but adsorption yields the optimal results in terms of energy cost, efficiency, and versatility with respect to the adsorbate [17]. Adsorption is one of the most efficient methods of removing VOCs, and various adsorbents are used, including carbon materials [18], silica [19], zeolites [20], and polymers [21]. Owing to their low costs and high stabilities, carbon materials are the most commonly used adsorbents in VOC removal [22]. Activated carbon, in particular, provides key advantages in adsorption because of its high porosity, large surface area, and rapid adsorption [23,24]. It is well-known that the intrinsic surface of activated carbon is hydrophobic. However, in many cases, the operation of activated carbon requires a hydrophilic environment such as water purification and wastewater treatment. Even VOC removal occurs in humid environments. Among the options of surface modification, acidic treatments are popular. Strong acids such as nitric acid are often used in surface modification. With the help of acidic treatment, the generation of carboxylic acid is noticeable along with hydroxyl groups. With the increasing need to remove specific chemicals such as aldehydes, the refined surface modification was necessitated with or without acidic treatment. Amination showed promising results in formaldehyde removal [18]. Previous studies have reported on the surface modification of activated carbon with amines that can be used to remove CH_3CHO from the atmosphere. The amine-containing materials used in previous studies include aminobenzenesulfonic acids, ethylenediamine, diethylenetriamine, and amino acids [25–27]. However, these materials are toxic and can have harmful effects on humans when the modified activated carbon is used in air-purifier filters. Therefore, developing non-toxic amine-modified activated carbon is essential, but no related research regarding the adsorption of CH_3CHO with non-toxic amines has been reported. In addition, the related references on acetaldehyde removal over activated carbon materials are summarized in Table 1.

In this study, we used a spherical bead-type activated carbon (BAC), which provides a pleasant working environment with no dust and high strength and fluidity. This type of BAC can be easily operated for adsorption and desorption, thereby making it feasible for recycling and helping for a circular economy. Specifically, compared to commercial granular activated carbon, BAC has a narrow-sized distribution. If BAC is applied to a BAC-incorporated membrane filter, a new design of the cartridge-type filter can be proposed. For a better understanding, the overall concept of BAC application is presented in Figure 1. Due to these properties, BAC is particularly useful as an air-purifier filter, and further research is necessary to explore its applicability [28]. The main aim of this study is to characterize the detailed adsorption behavior of CH_3CHO over BAC modified with piperazine, which is a non-toxic amine. Moreover, the reaction mechanism of CH_3CHO removal using the non-toxic amine-BAC is investigated in detail.

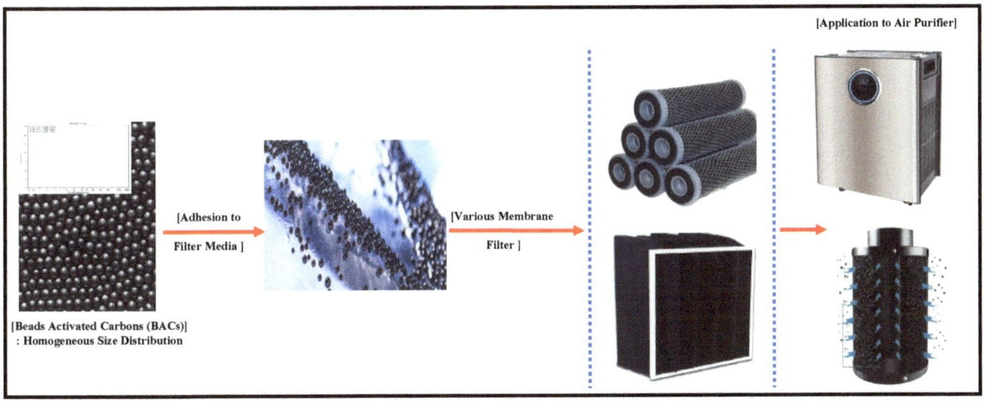

Figure 1. Conceptual illustration of BACs applied to air purification through systemizing membrane filtration.

Table 1. The related research on acetaldehyde removal using activated carbon.

References	Activated Carbon Type	Impregnated Material	Mechanism
[29]	AC (Calgon, Norit, and Westvaco)	nitric acid	(1) When very small pores as close as the size of the acetaldehyde molecule and oxygen-containing groups are present (to a certain extent) within AC, the heat of adsorption reaches its maximum value. (2) A low density of surface groups can enhance the heat of adsorption, whereas extensive oxidation leads to a decrease in the strength of adsorption forces. This happens due to the blockage of the pore entrances containing functional groups and the decrease in the accessibility of hydrophobic surface where the dispersive interactions of hydrocarbon moiety can be enhanced.
[30]	AC (Calgon and Westvaco)	urea (450/950 °C)	(1) The adsorption forces are strong in small pores, and their volume governs the adsorbed amount. (2) The absorbed amount can be enhanced when functional groups bearing nitrogen are present. (3) These groups can provide additional adsorption centers when the small pores are filled with acetaldehyde molecules.
[26]	AC (coconut-shell and coal-base)	amine	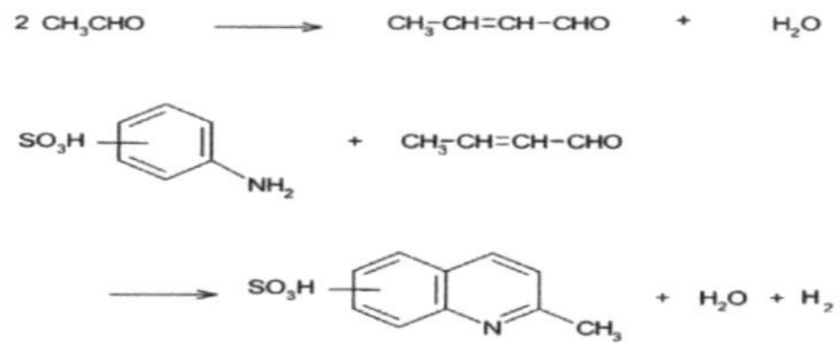
[31]	AC (corn grain)	KOH	(1) The effects of acetaldehyde adsorption on ACs were investigated in terms of textural properties, energetic heterogeneity, and surface chemistries. (2) The adsorption properties of water vapor were explained by the effect of the oxygen-containing groups on the surface of ACs over acetaldehyde adsorption. (3) The influences of pore size distribution (below 8 A°) and energetic heterogeneity of ACs on acetaldehyde adsorption were highly predominant compared to that of specific surface area and surface chemistry.
[32]	AC (coconut base)	-	The study established a semi-quantitative relationship between pore size distribution and energy in relation to adsorption kinetics; the wider and more heterogeneous porosities resulted in higher rate constants for the resin-based carbon when compared to the ultramicroporous nutshell material.
[33]	ACF	metal oxide	ACF-K-20/5%MgO revealed three types of surface adsorption sites: one was assigned to physisorption on the surface O-containing carbon groups and two other sites are placed on a MgO surface and provide acetaldehyde chemisorption in two different modes.
[34]	ACF (cellulose base)	aniline-ethanol	(1) $CH_3CHO_{(g)} \rightarrow CH_3CHO_{(AD)}$ [Adsorption] (2) $CH_3CHO_{(AD)} + \frac{1}{2}O_2 \rightarrow CH_3COOH$ [Oxidation] (3) $2CH_3COOH \rightarrow (CH_3CO)_2O + H_2O$ [Dehydration] (4) $(CH_3CO)_2O + C_6H_5NH_2 \rightarrow C_6H_5NCH_3CO - CH_3COOH$
[35]	ACF (HDPE fiber)	Ag	(1) Ag particles were precipitated on the surface of ACF through interactive affinity, and the carbonyl group of AA is in creased to show that AA is adsorbed on the AC surface. (2) The AA adsorption of ACF and Ag/ACF composites performed in this study was suitable for the dose–response model, and the experimental data showing the asymmetric shape of the AA adsorption breakthrough curve for ACF and Ag/ACF composites were satisfactorily fitted.
[36]	AC and ACF	amine	(1) The high BET surface area provides more sites for acetaldehyde adsorption. (2) ACF has a systematic open macrostructure, which drives a low-pressure drop and allows fast adsorption without diffu sion hindrance.

2. Experimental Section

2.1. Materials and Sample Preparation

BAC (particle diameter of <0.1 mm) was obtained from ZEOBuilder (Seoul, Republic of Korea) and heated at 900 °C for 3 h under N_2 flowing at 175 mL/min.

Piperazine, nitric acid, and ethyl alcohol were purchased from Samchun Pure Chemical (Pyeongtaek, Republic of Korea). Piperazine was prepared with ethyl alcohol, and nitric acid was prepared with distilled water. Aqueous stock solutions containing 1–10% (w/v) piperazine with or without 1% (w/v) nitric acid were prepared. The sample names differ according to the contents of the mixtures used (Table 2). The heated BAC was impregnated with the solution of piperazine with/without nitric acid, and the sample was then placed in a shaking water bath at 25 °C and shaken at 130 rpm for 24 h. After impregnation, the BAC was filtered and dried at 70 °C for 24 h. For comparison, piperazine was impregnated on coconut shell and coal-based activated carbons using the same procedure (Supplementary Information Figure S8).

Table 2. Description of each sample.

Sample	Piperazine % [w/v%]	Nitric Acid % [w/v%]	Heat Treatment Temp. [°C]
P1	1	-	-
P7	7	-	-
P7N1	7	1	-
P1N1-900	1	1	900
P3N1-900	3	1	900
P5N1-900	5	1	900
P7N1-900	7	1	900
P10N1-900	10	1	900

2.2. Methods

2.2.1. Preliminary Characterization of BACs

In order to investigate the characteristics of acetaldehyde removal for the BACs impregnated with piperazine/nitric acid, BARE-BAC was first examined by XRD, SEM, TEM, EDX, particle size distribution, as well as physical/chemical stability. All data is presented in the Supplementary Information (Figures S1–S7). In brief, BARE-BAC showed the amorphous phase (refer to XRD data in the Supplementary Information) without any crystallinity, therefore Raman characteristics were not necessary to verify its further crystallinity. BAC from a resin precursor contains a slight impurity of Si, characterized by SEM-EDX. Basically, approx. 1 wt% of impurity in activated carbon cannot affect its adsorption capacity. SEM morphology implies that the BACs were successfully synthesized from the resin precursor. The narrow window of particle size distribution in BARE-BAC ranged from 440 to 600 μm with a mean value of 512 μm, indicating homogeneous particle size compared with commercial granular activated carbon. It can be easily applied to the various type of membrane filter system. Moreover, the physical stability has been tested by attrition and abrasion (ASTM D4058-96 & SPENCE Method). Two tests were performed which showed the value of 99.77 with a deviation of 0.11. The acetaldehyde cycle test was carried out to confirm the chemical stability. The regeneration efficiency showed about 90% during 3 cycles of the adsorption–desorption process under the given operational conditions (200 ppm, GHSV: 37,500 h^{-1}, 25 °C).

2.2.2. N_2 Sorption

N_2 adsorption–desorption isotherms were obtained at 77 K using a surface area and pore size analyzer (BELSORP MAX G, Microtrac MRB, Osaka, Japan). Before the adsorption studies, the BAC was outgassed at 120 °C under vacuum for 8 h. The surface area was determined using the Brunauer–Emmett–Teller (BET) equation, the total pore volume was calculated at a relative pressure (P/P_0) of 0.99, and the micropore volume was calculated using the Horvath–Kawazoe (HK) and t-plot methods.

2.2.3. CHNS Elemental Analysis

A CHNS elemental analyzer (Flash 2000, Thermo Fisher Scientific, Waltham, MA, USA) with a thermal conductivity detector was used to examine the compositions of the BAC samples to determine the relative contents of C, H, N, and S as percentages.

2.2.4. Fourier Transform Infrared (FTIR) Spectroscopy

FTIR spectroscopy was used to qualitatively evaluate the chemical structures of the BAC samples. The FTIR spectra were measured using an FTIR spectrometer (US/iS50, Thermo Fisher Scientific) in the frequency range of 400–4000 cm^{-1}.

2.2.5. X-ray Photoelectron Spectroscopy (XPS)

XPS was performed using an XP spectrometer (K-Alpha$^+$, Thermo Fisher Scientific) with a monochromatic Al Kα (1486.6 eV) radiation source operated at 15 kW and 50 W. Prior to the analysis, the samples were outgassed at room temperature until the system pressure reached 5.2×10^{-9} torr. High-resolution spectra were collected at a constant pass energy of 29.35 eV over an area with a diameter of 4000 μm. The amount of each element (C, O, N, and S) was calculated using individual spectrum, and the energy scale was calibrated using the C 1s photoelectron line at 285.0 eV.

2.2.6. CH_3CHO Adsorption

Figure 2 schematically illustrates the continuous flow system employed in evaluating CH_3CHO removal using the BAC samples. Tests were conducted at 10 °C using a thermo-hygrostat (climatic chamber, Weiss Technik, Rieskirchen-Lindenstruth, Germany), and the BAC sample (2 mL) was placed in a stainless steel column. The CH_3CHO gas was 1000 ppm CH_3CHO in N_2, which was then mixed with air to yield a CH_3CHO concentration of 200 ppm. Gas analysis was performed using a sampling pump (GV-100S, Gastec, Ayase, Japan) and tube (92/92M/92L, Gastec). The breakthrough adsorption capacity of CH_3CHO (W_{ad} [mg/g]) was calculated by integrating the area under the breakthrough curve using the flow rate, CH_3CHO concentration, time, and adsorbent mass. All adsorption tests were stopped at $C_{out}/C_{in} = 1$ (C_{out} [ppm]: outlet concentration of CH_3CHO; C_{in} [ppm]: inlet concentration of CH_3CHO).

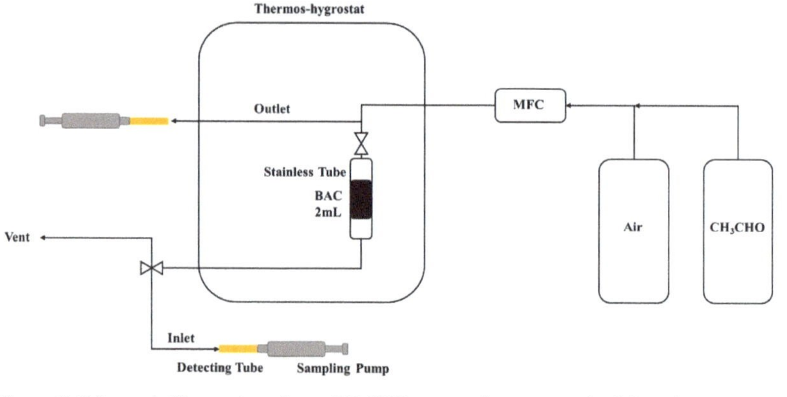

Figure 2. Schematic illustration of our CH_3CHO removal apparatus for lab-scale tests.

2.2.7. Thermal Regeneration

The modified BACs were thermally regenerated in a muffle furnace (Daihan Scientific, Wonju, Republic of Korea). After the CH_3CHO adsorption studies, the spent BAC samples (approximately 2 mL) were heated from room temperature to 453 K (heating rate: 5 K/min) and maintained at this temperature for 2 h in air.

3. Results

3.1. Characterizations of BAC

3.1.1. Textural Structure

The textural properties of the BAC samples, which were determined using the N_2 adsorption–desorption isotherms, are summarized in Table 3. The BET surface areas (S_{BET}) decrease in the following order: BARE-BAC > piperazine/nitric acid-co-impregnated BACs > piperazine-modified BACs. Piperazine treatment decreases the S_{BET} of BARE-BAC, but the co-impregnation with piperazine and nitric acid minimizes the decrease, likely because the nitric acid molecules adsorbed on the pore surfaces continue to penetrate the pore walls during impregnation. However, as the piperazine content increases, the pores of the BAC are blocked, and S_{BET} decreases. BARE-BAC displays the highest S_{BET} (1442.1 m^2/g) and the highest micropore and total pore volumes (0.6189 and 0.6284 cm^3/g, respectively), which are almost 6–77% higher than those of the modified BACs. Therefore, chemical factors have a greater effect on CH_3CHO adsorption than physical factors.

Table 3. Textural properties of BAC samples.

Sample	S_{BET} [m^2/g]	S_{Micro} [m^2/g]	V_{Total} [cm^3/g]	V_{Micro} [cm^3/g]	Average Pore Diameter [nm]
BARE-BAC	1442.1	1437.3	0.6284	0.6189	1.7429
P1	921.5	916.9	0.4123	0.4033	1.7898
P7	794.5	791.0	0.3533	0.3462	1.7788
P7N1	1141.3	1137.1	0.5001	0.4916	1.7528
P1N1-900	1347.2	1341.2	0.5905	0.5788	1.7533
P3N1-900	1275.3	1270.2	0.5612	0.5508	1.7602
P5N1-900	1191.6	1185.8	0.5259	0.5142	1.7652
P7N1-900	1115.3	1110.0	0.4818	0.4711	1.7281
P10N1-900	983.8	979.3	0.4390	0.4298	1.7850

The piperazine/nitric acid-co-impregnated BACs exhibit larger micropore volumes than the piperazine-modified BACs. Overall, co-impregnation with piperazine and nitric acid should result in different modifications of the BAC surface compared to those caused by piperazine modification of the BAC.

3.1.2. CHNS Elemental Analysis

The contents of C, H, N, and S in the samples were measured using a CHNS elemental analyzer, and the results are shown in Table 4. The N and H contents increase after treatment with piperazine and nitric acid, respectively, and the introduced amine (–NH) groups are critical in CH_3CHO removal via chemisorption [26,27,34].

Table 4. Contents of C, H, N, and S in BAC samples.

Sample	C [%]	H [%]	N [%]	S [%]
BARE-BAC	93.84	0.45	* ND	1.31
P1	80.18	2.16	0.63	1.01
P7	87.00	1.64	3.52	1.13
P7N1	85.90	1.74	1.89	1.21
P1N1-900	83.02	2.30	0.39	0.97
P3N1-900	93.39	0.79	1.23	1.25
P5N1-900	81.99	2.67	1.52	0.94
P7N1-900	83.21	2.21	2.13	0.98
P10N1-900	80.89	2.77	2.67	0.89

* ND: No data.

3.1.3. Chemical Characterization

Figure 3 shows the FTIR spectra of BARE-BAC and P7N1-900, which adsorbs the highest amount of CH_3CHO. The characteristic peaks attributed to the functional groups are almost identical in the two spectra. The O–H and N–H bonds result in noticeable peaks, and Table 5 summarizes the intensity data of the functional groups. The spectrum of P7N1-900 displays peaks with stronger intensities the O–H and amine functional groups representing treatment with piperazine and nitric acid. Penchah et al. [37] reported that nitric acid increases the carboxylic acid (COOH) content on the surface of activated carbon, and piperazine is linked to COOH. Therefore, the number of OH groups of P7N1-900 is increased approximately 2-fold on account of using nitric acid, and the number of amine groups is increased approximately 4.2-fold on account of using piperazine. In addition, no change in the structure of BAC is observed, as there is almost no change in the $-CH_2$ groups.

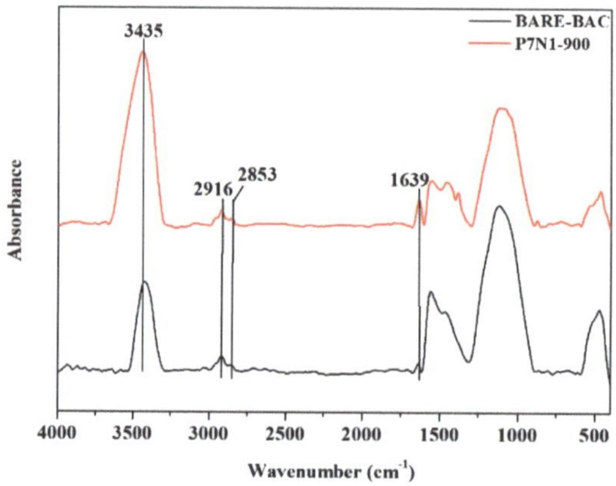

Figure 3. FTIR spectrum of BARE-BAC and P7N1-900.

Table 5. Intensity of BARE-BAC and P7N1-900's functional groups.

Band Position [cm^{-1}]	Component	Intensity	
		BARE-BAC	P7N1-900
3435	O-H	3.40	6.95
2916	Saturated aliphatic CH_2	0.42	0.46
2853	Saturated aliphatic CH_2	0.16	0.21
1639	Amie, primary/secondary NH	0.29	1.23

Sources: Adapted from [38].

Ryu et al. [34] reported that the amine functional groups on activated carbon lead to a higher CH_3CHO adsorption efficiency, and thus, the amine groups should increase CH_3CHO adsorption.

3.1.4. XPS

XPS was used to investigate the chemical states of the elements in the BAC samples. The deconvoluted spectra are shown in Figure 4, and the relative abundances of C, O, N, and S are presented in Table 6.

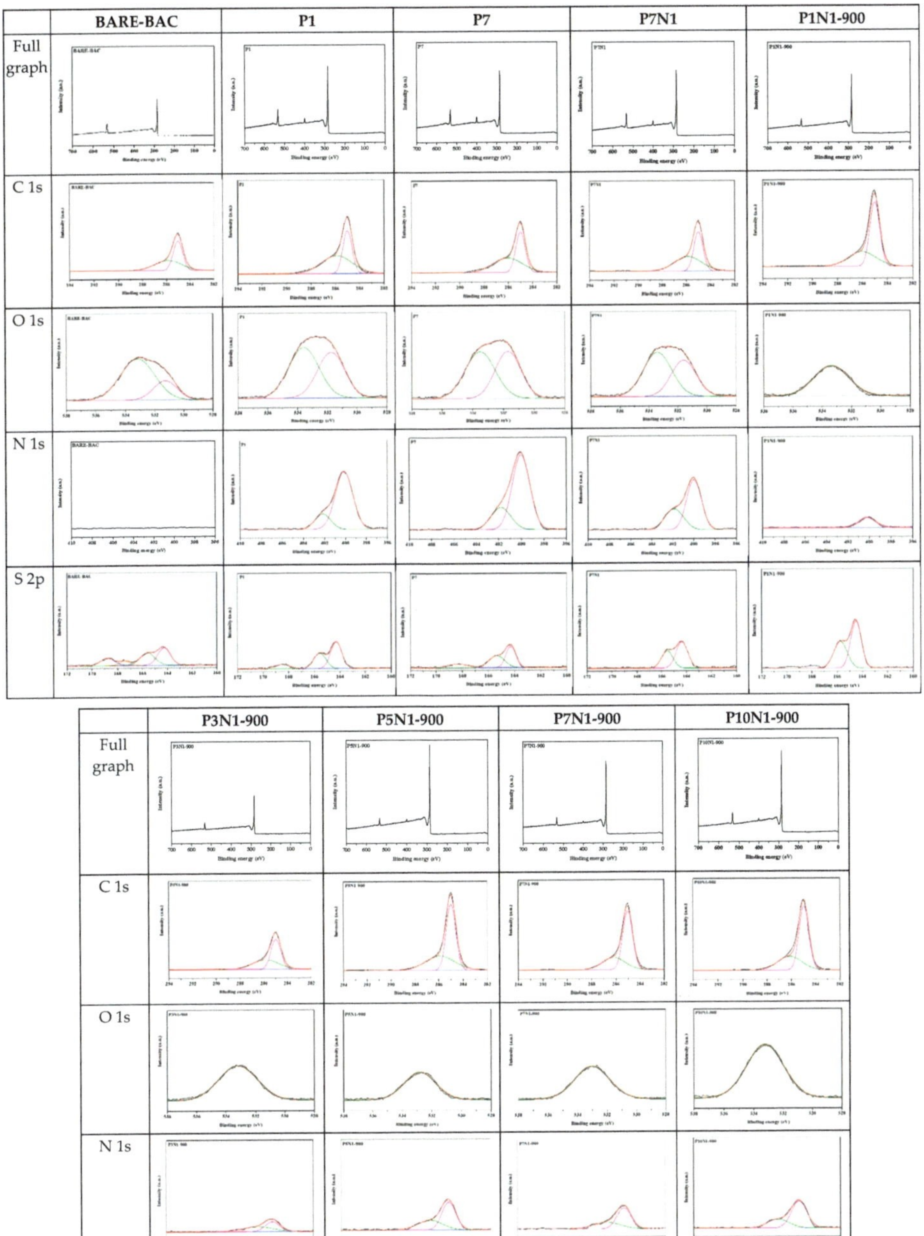

Figure 4. High resolution XPS spectra of C 1s, O 1s, N 1s, and S 2p for all BAC samples. (black line: experimental value, red line: fitting value).

After modification with piperazine, the number of N bonds increases, whereas the number of bonds involving S decreases. After heating and modification with piperazine and nitric acid, the number of N bonds increases, and the number of bonds involving O and S decreases.

The various N components in BAC were further determined by fitting the N 1s spectra. As shown in Figure 4 and Table 5, deconvolution of the N 1s spectrum reveals the presence of N–$(C)_3$ (tertiary nitrogen, secondary amine) and C–N^+O–C (oxidized nitrogen functionalities). After piperazine treatment, the number of assigned N bonds increases, particularly the N–$(C)_3$ content, and thus, when only piperazine is used, it coats the BAC surface. By contrast, in the BAC treated with piperazine and nitric acid, the content of C–N^+O–C increases by >50% and then decreases as the piperazine content increases. Thus, COOH groups are formed on the surface of the BAC via the addition of nitric acid, and piperazine is linked to these groups. The piperazine content of P1N1-900 is excessively low to link to the COOH groups, and thus, the content of N–$(C)_3$ is 100%. Conversely, in P3N1-900, piperazine and the COOH groups are linked, increasing the content of C–N^+O–C. In P5N1-900, the COOH groups and piperazine are linked, the remaining piperazine is coated on the surface of the BAC, and the content of N–$(C)_3$ is increased. N atoms doped onto the BAC surface may react with CH_3CHO to enhance its adsorption [30].

As shown in Figure 4 and Table 5, deconvolution of the O 1s and S 2p spectra reveal the presence of O–C/O–S (in phenol/epoxy or thioether/sulfonic acid), O=C/O=S (in carboxy/carbonyl or sulfoxides/sulfones), C–S–C (in sulfides), R–S–S–OR (in thioethers), R_2–S=O (in sulfoxides), and R–SO_2–R (in sulfones). During heating, the high temperature may convert O=C/O=S to O–C/O–S, and R_2–S=O and R–SO_2–R in P1N1-900–P10N1-900 are converted to C–S–C via heating. S doped on the BAC surface may also react with aldehydes to increase the HCHO adsorption performance [39].

Table 6. Results of the deconvolution of the XPS spectra of C 1s, O 1s, N 1s, and S 2p (in bold—atomic % of specific elements).

Bond Assignment	Energy [eV]	BARE-BAC [%]	P1 [%]	P7 [%]	P7N1 [%]	P1N1-900 [%]	P3N1-900 [%]	P5N1-900 [%]	P7N1-900 [%]	P10N1-900 [%]
C 1s										
C-C sp^2	284.8	51.15	41.23	46.46	47.57	62.49	55.21	60.32	69.02	66.84
C-O (phenol, alcohol, ether), C=N (amine, amide)	286.0–286.3	48.85	58.77	53.54	52.43	37.51	44.79	39.68	30.98	33.16
O 1s										
O-C/O-S (in phenol/epoxy or thioethers/sulfonic)	533.3–533.6	77.78	55.29	54.16	59.19	100	100	100	100	100
O=C/O=S (in carboxy/carbonyl or sulfoxides/sulfones)	532.0–532.5	22.22	44.71	45.84	40.81	-	-	-	-	-
N 1s										
N-$(C)_3$ (tertiary nitrogen, secondary amine)	399.1–400.0	-	83.51	76.46	68.18	100	100	63.78	67.99	70.61
C-N^+O-C (oxidized nitrogen functionalities)	402.3	-	16.49	23.54	31.82	-	-	36.22	32.01	29.39
S 2p										
C-S-C (in sulfides); R-S-S-OR (in thioethers)	164.5–166.0	81.74	96.35	93.56	93.56	100	100	100	100	100
R_2-S=O (in sulfoxides)	167.0–167.3	2.28	-	-	-	-	-	-	-	-
R-SO_2-R (in sulfones)	168.4–168.6	15.98	3.65	6.44	6.44	-	-	-	-	-

Sources: Adapted from [40].

3.2. CH_3CHO Adsorption

Figure 5 shows the CH_3CHO breakthrough curves measured at 10 °C in dry conditions, and the calculated adsorption parameters are listed in Table 6. The modified BAC samples display higher W_{ad} values, as shown in Table 7, than that of the unmodified BAC: 17.17, 18.73, 50.05, 57.44, 18.84, 54.48, 62.95, 72.34, and 61.38 mg/g for BARE-BAC, P1, P7, P7N1, P1N1-900, P3N1-900, P5N1-900, P7N1-900, and P10N1-900, respectively. The samples co-impregnated with piperazine and nitric acid generally exhibit higher W_{ad} values than those of the samples treated only with piperazine (P1 and P7). P7N1-900, in particular, adsorbs the largest amount of CH_3CHO, which is 4.2-fold higher than that adsorbed by BARE-BAC. Physical factors, such as the high S_{BET} values of the modified BAC samples and abundant micropores, likely contribute to the adsorption performances of the co-impregnated BAC samples. Moreover, heating destroys the functional groups on the surface of the BAC and ensures that the active sites comprising piperazine and nitric acid are evenly distributed. As the piperazine content increases, the amount of adsorbed CH_3CHO increases, but when the piperazine content is excessive, as in P10N1-900, the pores of the BAC are blocked and the adsorption performance is reduced. Chemical factors, such as the addition of nitric acid, result in higher CH_3CHO adsorption. The chemical reactions between CH_3CHO and linked piperazine explain the increased CH_3CHO adsorption levels of the BAC samples co-impregnated with piperazine and nitric acid compared to those of the other samples. This is corroborated by the XPS spectra, which show that CH_3CHO reacts strongly with C–N$^+$O–C (oxidized nitrogen functionalities) and N–$(C)_3$ (tertiary nitrogen, secondary amine). The reaction with CH_3CHO is stronger when both types of bonds are present compared to that when one type is present. Figure 6 shows the comprehensive mechanisms of amine modification of the BAC and CH_3CHO adsorption on the modified BAC surface.

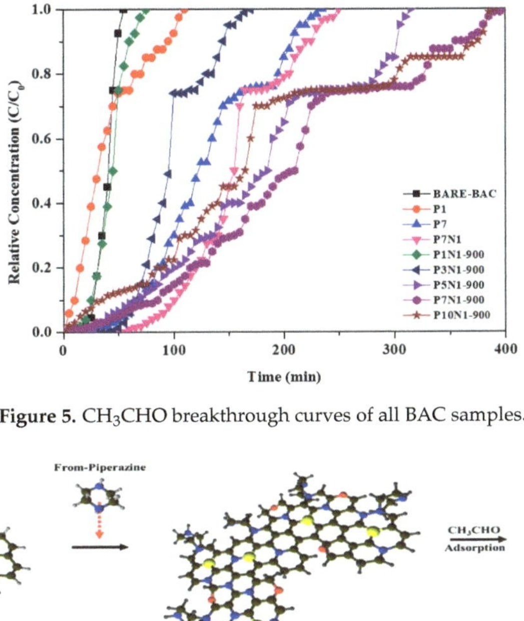

Figure 5. CH_3CHO breakthrough curves of all BAC samples.

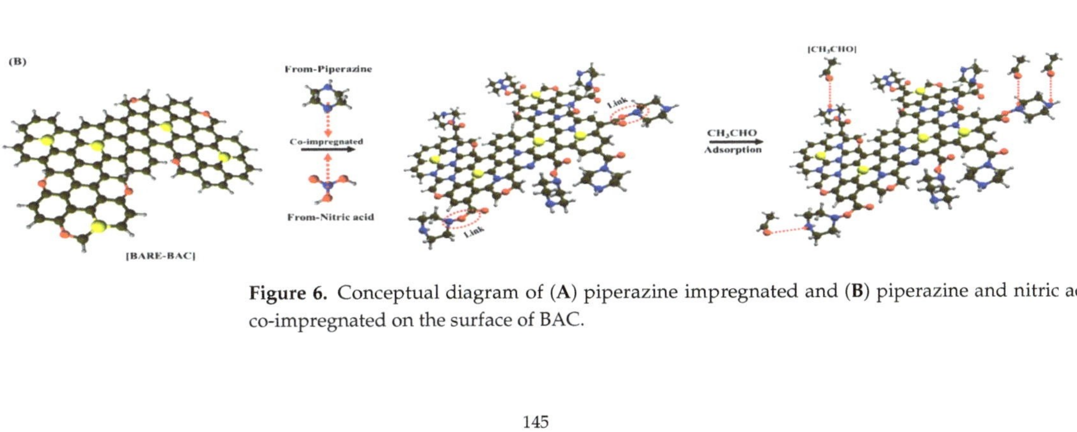

Figure 6. Conceptual diagram of (**A**) piperazine impregnated and (**B**) piperazine and nitric acid co-impregnated on the surface of BAC.

Table 7. CH_3CHO adsorption parameters on all BAC samples.

Sample	C_{in} [ppm]	W_{ad} [mg/g]
BARE-BAC	200	17.17
P1	200	18.73
P7	200	50.05
P7N1	200	57.44
P1N1-900	200	18.84
P3N1-900	200	54.48
P5N1-900	200	62.95
P7N1-900	200	72.34
P10N1-900	200	61.38

3.3. *Effect of Thermal Regeneration on CH_3CHO Adsorption*

Finally, we evaluated the thermal regeneration efficiencies of BARE-BAC and P7N1-900 after CH_3CHO adsorption. The BAC samples were subjected to three cycles of CH_3CHO adsorption and regeneration. Thermal regeneration was performed by heating the samples at 5 K/min to 453 K in a muffle furnace and maintaining this temperature for 2 h in air. Figure 7 shows the breakthrough curves obtained under dry conditions, and Table 8 presents the data. The results of Cycles 2 and 3 exhibit similar trends to those shown in Figure 5. As shown in Table 8, the respective regeneration efficiencies of BARE-BAC and P7N1-900 are 86.78% and 8.77%. Amine modification significantly reduces the regeneration efficiency of BARE-BAC by a factor of 10. BARE-BAC exhibits a high ratio of physical adsorption, whereas P7N1-900 displays a high ratio of chemical adsorption. BARE-BAC physisorbs CH_3CHO and desorbs it well in all cycles. By contrast, P7N1-900 chemisorbs CH_3CHO in Cycle 1 but does not desorb well, and thus, the CH_3CHO adsorption declines significantly in Cycle 2. Meanwhile, the amounts of CH_3CHO adsorbed in Cycles 2 and 3 are similar.

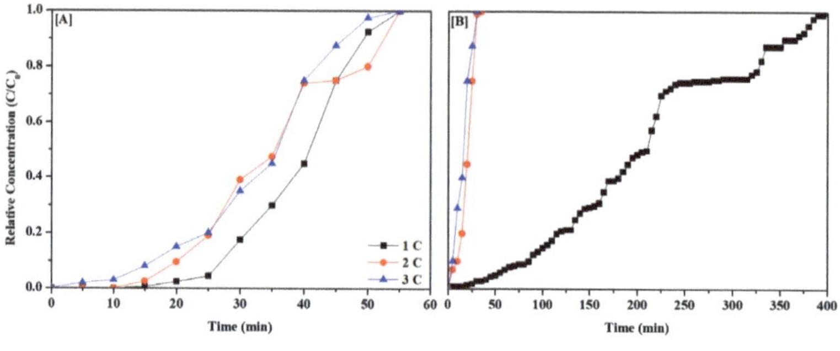

Figure 7. CH_3CHO regeneration efficiency of 3 cycles: BARE-BAC (**A**) and P7N1-900 (**B**).

Table 8. CH_3CHO regeneration efficiency of three cycles.

Sample	Number of Cycle	Adsorption Amount [mg/g]	Regeneration Efficiency of 3 Cycles [%]
BARE-BAC	1	17.17	86.78
	2	15.89	
	3	14.90	
P7N1-900	1	72.34	8.77
	2	6.13	
	3	5.97	

4. Conclusions

Customized BAC was heated at 900 °C and impregnated with piperazine or binary piperazine/nitric acid to enhance CH_3CHO removal via adsorption. The amount of piperazine was varied from 1 to 10% (*w*/*v*) and nitric acid was used at 1% (*w*/*v*). When only piperazine was used, the pore and micropore volumes of the modified BACs decreased with increasing piperazine content. The S_{BET} of BARE-BAC was 1442.1 m^2/g, and that of P7 was reduced to 794.5 m^2/g (modified BAC with 7 % *w*/*v* impregnation of piperazine). Meanwhile, co-impregnation resulted in a minimized decrease in S_{BET}, e.g., 1347.2 m^2/g for P1N1-900 (1% *w/v* each of piperazine and nitric acid, followed by heating at 900 °C). This suggests that nitric acid penetrates into the BAC microstructure to form pores of larger size and volume. However, the qualitative chemical structures of BARE-BAC and P7N1-900 (impregnation with 7% *w/v* piperazine and 1% *w/v* nitric acid, followed by heating at 900 °C) were not significantly different according to FTIR spectroscopy. Elemental analysis revealed that amine treatment increased the N contents of the BAC samples. When CH_3CHO adsorption was performed at 10 °C under dry conditions, the adsorption performance of P7N1-900 increased 4.2-fold compared to that of BARE-BAC. This suggests that the chemical factors of BAC have a greater effect on CH_3CHO adsorption than the physical factors. In regard to the impregnation with piperazine only (P1 and P7), the lower CH_3CHO adsorption capacities could be explained by linkages to hidden amines and intrinsic COOH groups, resulting in negative effects on adsorption. By contrast, P1N1-900 and P7N1-900 should exhibit linked piperazine-amines but still contain carboxylic acids on the surfaces of the BAC samples, resulting in increased levels of CH_3CHO adsorption. The CH_3CHO adsorption capacity of BARE-BAC was 17.17 mg/g, and those of P7 and P7N1-900 increased to 50.05 and 72.34 mg/g, respectively. The XP spectra indicated that N–$(C)_3$ (tertiary nitrogen, secondary amine) and C–N^+O–C (oxidized nitrogen functionalities) were present throughout the amine-treated BAC samples. The presence of C–N^+O–C indirectly indicated that the amines were linked to COOH groups, which were mostly generated via nitric acid treatment. An increased amine content resulted in higher concentrations of N–$(C)_3$, and thus, amine treatment introduced primary and secondary NH functional groups into the modified BAC and facilitated CH_3CHO adsorption. Although CH_3CHO adsorption was only evaluated under flow conditions, the modified BAC samples should display high CH_3CHO adsorption capacities even under static conditions.

Supplementary Materials: The following supporting information can be downloaded at: https://www.mdpi.com/article/10.3390/membranes13060595/s1, Figure S1: SEM Morphology of Bare BAC, Figure S2: TEM Morphology of Bare BAC (After Grinding/Crushing BAC), Figure S3: XRD Patterns of Bare BAC, Figure S4: SEM-EDX of Bare BAC, Figure S5: Attrition and Abrasion Test of Bare BAC (Physical Stability), Figure S6: Acetaldehyde Cycle Test over Bare BAC, Figure S7: Particle Size Distribution of Bare BAC, Figure S8: CH_2CHO Adsorption Parameters on Cucunut Shell and Coal Based Activated Carbon Samples.

Author Contributions: Conceptualization, Y.-J.K. (Yu-Jin Kang) and S.-J.Y.; methodology, Y.-J.K. (Yu-Jin Kang), D.-J.S., Y.-J.K. (Yu-Jin Kim) and H.-R.C.; validation, Y.-J.K. (Yu-Jin Kim) and K.O.; investigation, Y.-J.K. (Yu-Jin Kang), D.-J.S., Y.-J.K. (Yu-Jin Kim) and H.-R.C.; resources, S.-H.Y. and J.-I.P.; data curation, Y.-J.K. (Yu-Jin Kang) and K.O.; writing—original draft preparation, Y.-J.K. (Yu-Jin Kang); writing—review and editing, Y.-J.K. (Yu-Jin Kang) and Y.-J.K. (Yu-Jin Kim); visualization, Y.-J.K. (Yu-Jin Kang); supervision, S.-H.Y. and J.-I.P.; project administration, S.-H.Y. and J.-I.P. All authors have read and agreed to the published version of the manuscript.

Funding: This research was supported by the KNUDP (Republic of Korea National University Development Project) funded by the Ministry of Education (MOE, Republic of Korea) and National Research Foundation of Republic of Korea (NRF).

Institutional Review Board Statement: Not applicable.

Data Availability Statement: The data presented in this study are available upon request from the corresponding author.

Conflicts of Interest: The authors declare no conflict of interest.

References

1. Lee, H.W.; Kim, S.; Ryu, C.; Park, S.H.; Park, Y.K. Review of the use of activated biochar for energy and environmental applications. *Carbon Lett.* **2018**, *26*, 1–10.
2. Isinkaralar, K.; Turkyilmaz, A. Simultaneous adsorption of selected VOCs in the gas environment by low-cost adsorbent from Ricinus communis. *Carbon Lett.* **2022**, *32*, 1781–1789. [CrossRef]
3. Ryu, D.Y.; Kim, D.W.; Kang, Y.J.; Lee, Y.J.; Nakabayashi, K.; Miyawaki, J.; Park, J.I.; Yoon, S.H. Preparation of environmental-friendly N-rich chitin-derived activated carbon for the removal of formaldehyde. *Carbon Lett.* **2022**, *32*, 1473–1479. [CrossRef]
4. Lee, Y.J.; Lee, H.M.; Kim, B.J. Facile microwave treatment of activated carbons and its effects on hydrocarbon adsorption/desorption behaviors. *Carbon Lett.* **2023**, *33*, 1–10. [CrossRef]
5. Diana, P.; Alexander, O.; Tikhon, K.; Eugenii, G.; Elena, I.; Alexander, Z. Carbonization of oriented poly(vinyl alcohol) fibers impregnated with potassium bisulfate. *Carbon Lett.* **2020**, *30*, 637–650.
6. Azuma, K.; Uchiyama, I.; Uchiyama, S.; Kunugita, N. Assessment of inhalation exposure to indoor air pollutants: Screening for health risks of multiple pollutants in Japanese dwellings. *Environ. Res.* **2016**, *145*, 39–49. [CrossRef]
7. Ghaffarianhoseini, A.; AlWaer, H.; Omrany, H.; Ghaffarianhoseini, A.; Alalouch, C.; Clements-Croome, D.; Tookey, J. Sick building syndrome: Are we doing enough? *Archit. Sci. Rev.* **2018**, *61*, 99–121. [CrossRef]
8. Jansz, J. Theories and knowledge about sick building syndrome. *SBS* **2011**, *591*, 25–58.
9. Naddafi, K.; Nabizadeh, R.; Rostami, R.; Ghaffari, H.R.; Fazlzadeh, M. Formaldehyde and acetaldehyde in the indoor air of waterpipe cafés: Measuring exposures and assessing health effects. *Build. Environ.* **2019**, *165*, 106392. [CrossRef]
10. World Health Organization. Acetaldehyde: Health and safety guide. *Health Saf. Guide* **1994**, *90*, 32.
11. Nikawa, T.; Naya, S.I.; Tada, H. Rapid removal and decomposition of gaseous acetaldehyde by the thermo-and photo-catalysis of gold nanoparticle-loaded anatase titanium (IV) oxide. *J. Colloid Interface Sci.* **2015**, *456*, 161–165. [CrossRef]
12. Gabelman, A.; Hwang, S.T. Hollow fiber membrane contactors. *J. Membr. Sci.* **1999**, *159*, 61–106. [CrossRef]
13. Busca, G.; Berardinelli, S.; Resini, C.; Arrighi, L. Technologies for the removal of phenol from fluid streams: A short review of recent developments. *J. Hazard. Mater.* **2008**, *160*, 265–288. [CrossRef] [PubMed]
14. Kennes, C.; Thalasso, F. Waste gas biotreatment technology. *J. Chem. Technol. Biotechnol.* **1998**, *72*, 303–319. [CrossRef]
15. Lahousse, C.; Bernier, A.; Grange, P.; Delmon, B.; Papaefthimiou, P.; Ioannides, T.; Verykios, X. Evaluation of γ-MnO_2 as a VOC removal catalyst: Comparison with a noble metal catalyst. *J. Catal.* **1998**, *178*, 214–225. [CrossRef]
16. Parmar, G.R.; Rao, N.N. Emerging control technologies for volatile organic compounds. *Crit. Rev. Environ. Sci. Technol.* **2008**, *39*, 41–78. [CrossRef]
17. Gil, E.R.; Ruiz, B.; Lozano, M.S.; Martín, M.J.; Fuente, E. VOCs removal by adsorption onto activated carbons from biocollagenic wastes of vegetable tanning. *Chem. Eng. J.* **2014**, *245*, 80–88. [CrossRef]
18. Kang, Y.J.; Jo, H.K.; Jang, M.H.; Ma, X.; Jeon, Y.; Oh, K.; Park, J.I. A Brief Review of Formaldehyde Removal through Activated Carbon Adsorption. *Appl. Sci.* **2022**, *12*, 5025. [CrossRef]
19. Zaitan, H.; Manero, M.H.; Valdés, H. Application of high silica zeolite ZSM-5 in a hybrid treatment process based on sequential adsorption and ozonation for VOCs elimination. *J. Environ. Sci.* **2016**, *41*, 59–68. [CrossRef]
20. Kim, K.J.; Ahn, H.G. The effect of pore structure of zeolite on the adsorption of VOCs and their desorption properties by microwave heating. *Microporous Mesoporous Mater.* **2012**, *152*, 78–83. [CrossRef]
21. Liu, H.; Yu, Y.; Shao, Q.; Long, C. Porous polymeric resin for adsorbing low concentration of VOCs: Unveiling adsorption mechanism and effect of VOCs' molecular properties. *Sep. Purif. Technol.* **2019**, *228*, 115755. [CrossRef]
22. Bradley, R.H. Recent developments in the physical adsorption of toxic organic vapours by activated carbons. *Adsorpt. Sci. Technol.* **2011**, *29*, 1–28. [CrossRef]
23. Das, D.; Gaur, V.; Verma, N. Removal of volatile organic compound by activated carbon fiber. *Carbon* **2004**, *42*, 2949–2962. [CrossRef]
24. Singh, K.P.; Mohan, D.; Tandon, G.S.; Gupta, G.S.D. Vapor-phase adsorption of hexane and benzene on activated carbon fabric cloth: Equilibria and rate studies. *Ind. Eng. Chem.* **2002**, *41*, 2480–2486. [CrossRef]
25. Hayashi, T.; Kumita, M.; Otani, Y. Removal of acetaldehyde vapor with impregnated activated Carbons: Effects of steric structure on impregnant and acidity. *Environ. Sci. Technol.* **2005**, *39*, 5436–5441. [CrossRef]
26. Vikrant, K.; Qu, Y.; Kim, K.H.; Boukhvalov, D.W.; Ahn, W.S. Amine-functionalized microporous covalent organic polymers for adsorptive removal of a gaseous aliphatic aldehyde mixture. *Environ. Sci. Nano* **2020**, *7*, 3447–3468. [CrossRef]
27. Yamashita, K.; Noguchi, M.; Mizukoshi, A.; Yanagisawa, Y. Acetaldehyde removal from indoor air through chemical absorption using L-cysteine. *Int. J. Environ. Res. Public Health* **2010**, *7*, 3489–3498. [CrossRef] [PubMed]
28. Son, B.C.; Park, C.H.; Kim, C.S. Fabrication of antimicrobial nanofiber air filter using activated carbon and cinnamon essential oil. *J. Nanosci. Nanotechnol.* **2020**, *20*, 4376–4380. [CrossRef] [PubMed]
29. El-Sayed, Y.; Bandosz, T.J. A study of acetaldehyde adsorption on activated carbons. *J. Colloid Interface Sci.* **2001**, *242*, 44–51. [CrossRef]
30. El-Sayed, Y.; Bandosz, T.J. Acetaldehyde adsorption on nitrogen-containing activated carbons. *Langmuir* **2002**, *18*, 3213–3218. [CrossRef]

31. Park, K.H.; Shim, W.G.; Shon, H.K.; Lee, S.G.; Ngo, H.H.; Vigneswaran, S.; Moon, H. Adsorption characteristics of acetaldehyde on activated carbons prepared from corn-based biomass precursor. *Sep. Sci. Technol.* **2010**, *45*, 1084–1091. [CrossRef]
32. Branton, P.; Bradley, R.H. Effects of active carbon pore size distributions on adsorption of toxic organic compounds. *Adsorption* **2011**, *17*, 293–301. [CrossRef]
33. Baur, G.B.; Yuranov, I.; Kiwi-Minsker, L. Activated carbon fibers modified by metal oxide as effective structured adsorbents for acetaldehyde. *Catal. Today* **2015**, *249*, 252–258. [CrossRef]
34. Ryu, D.Y.; Nakabayashi, K.; Shimohara, T.; Morio, U.; Mochida, I.; Miyawaki, J.; Jeon, Y.; Park, J.I.; Yoon, S.H. Behaviors of cellulose-based activated carbon fiber for acetaldehyde adsorption at low concentration. *Appl. Sci.* **2019**, *10*, 25. [CrossRef]
35. Kim, B.J.; An, K.H.; Shim, W.G.; Park, Y.K.; Park, J.; Lee, H.; Jung, S.C. Acetaldehyde Adsorption Characteristics of Ag/ACF Composite Prepared by Liquid Phase Plasma Method. *Nano Mater.* **2021**, *11*, 2344. [CrossRef] [PubMed]
36. Park, S.; Yaqub, M.; Lee, S.; Lee, W. Adsorption of acetaldehyde from air by activated carbon and carbon fibers. *Environ. Eng. Res.* **2022**, *27*, 2. [CrossRef]
37. Ramezanipour Penchah, H.; Ghaemi, A.; Jafari, F. Piperazine-modified activated carbon as a novel adsorbent for CO_2 capture: Modeling and characterization. *Environ. Sci. Pollut. Res.* **2022**, *29*, 5134–5143. [CrossRef]
38. Karnati, S.R.; Høgsaa, B.; Zhang, L.; Fini, E.H. Developing carbon nanoparticles with tunable morphology and surface chemistry for use in construction. *Constr. Build. Mater.* **2020**, *262*, 120780. [CrossRef]
39. Kang, Y.J.; Jo, H.K.; Jang, M.J.; Han, G.J.; Yoon, S.J.; Oh, K.; Park, J.I. Acid treatment enhances performance of beads activated carbon for formaldehyde removal. *Carbon Lett.* **2023**, *33*, 397–408. [CrossRef]
40. Seredych, M.; Łoś, S.; Giannakoudakis, D.A.; Rodríguez-Castellón, E.; Bandosz, T.J. Photoactivity of g-C3N4 /S-Doped Porous Carbon Composite: Synergistic Effect of Composite Formation. *ChemSusChem* **2016**, *9*, 795–799. [CrossRef]

membranes

MDPI

Article

Purification of Liquid Fraction of Digestates from Different Origins—Comparison of Polymeric and Ceramic Ultrafiltration Membranes Used for This Purpose

Agnieszka Urbanowska

Department of Water, Wastewater and Waste Technology, Faculty of Environmental Engineering, Wroclaw University of Science and Technology, Wybrzeze Wyspianskiego 27, 50-370 Wroclaw, Poland; agnieszka.urbanowska@pwr.edu.pl

Abstract: Circular economy, clean technologies, and renewable energy are key to climate protection and modern environmental technology. Recovering water and valuable minerals from the liquid fraction of digestate is in line with this strategy. Digestate, a byproduct of anaerobic methane fermentation in biogas plants, is a potential source of water, minerals for fertilizers, and energy rather than waste. This study examined digestate from municipal and agricultural biogas plants and highlights the need for research on both due to their differences. The use of membrane techniques for water recovery from liquid digestate offers an innovative alternative to conventional methods. This study used standalone membrane filtration and an integrated system to produce water suitable for agricultural use. Ceramic membranes with cut-offs of 1, 5, 15, and 50 kDa and polymeric membranes of polyethersulfone and regenerated cellulose with cut-offs of 10 and 30 kDa were tested. The results showed that the membrane material significantly affects the transport and separation properties. Higher cut-off values increased permeate flux across all membranes. Ceramic membranes were more susceptible to fouling in standalone ultrafiltration, but were more effective in purifying digestate than polymeric membranes. The best results were obtained with a ceramic membrane with a 1 kDa cut-off (for example, for the integrated process and the municipal digestate, the retention rates of COD, BOD_5 and DOC were 69%, 62%, and 75%, respectively).

Keywords: digestate; municipal waste biogas plant; agricultural biogas plant; membrane processes; polymeric membranes; ceramic membranes; integrated processes

Citation: Urbanowska, A. Purification of Liquid Fraction of Digestates from Different Origins—Comparison of Polymeric and Ceramic Ultrafiltration Membranes Used for This Purpose. *Membranes* **2024**, *14*, 203. https://doi.org/10.3390/membranes14100203

Academic Editors: Vitaly Gitis and Konstantinos Beltsios

Received: 15 August 2024
Revised: 8 September 2024
Accepted: 23 September 2024
Published: 25 September 2024

1. Introduction

The combined effects of demographic growth, the evolving lifestyle of the consumer, economic and industrial development, and climate change have resulted in significant challenges to ensure sufficient and sustainable water resources for municipal and industrial use, as well as agricultural production in many regions of the world [1]. This places a significant burden on present and future generations with regard to the responsibility for the natural environment and the restoration of the ecological balance [2]. The current limitation in the availability of nonrenewable raw material resources is causing a change in environmental protection strategies towards the development of so-called clean production [3]. This can be achieved, for instance, by reducing the amount of pollutants generated or eliminating them entirely, and by producing reusable products. According to the principles of clean production, waste should be transformed into a substrate for a new full-value product [4]. In municipal and agricultural biogas plants, the underlying principles include the use of the methane fermentation process to generate green energy (biogas) and the optimization of the recovery of nutrients from the byproducts of this process [5]. Digestate, the byproduct of the fermentation process in both types of biogas plants, represents a significant environmental concern [6]. The digestate primarily comprises undigested organic matter and mineral components. The aforementioned components are present in quantities comparable to their content in the substrates used in the biogas plant [7]. The quantity of digestate produced is estimated to be in the range of 85–95% of the weight of the substrates employed [8]. The proportion of digestate produced is inversely proportional to the degree of fermentation of the organic matter present in the raw material [9]. The quality of digestate also depends on the source of the microbial inoculum, the conditions under which the fermentation process occurred (pH, temperature, organic matter loading rate, hydraulic retention time), and the type of biogas plant and the type of technology used [10]. Furthermore, pretreatment of input components also influences the physicochemical composition of the final byproduct [11].

It has been incumbent upon municipal facilities to manage and process the digestate primarily through dewatering, drying, and thermal incineration [12]. In contrast, agricultural biogas plants are faced with the challenge of excess digestate production and inadequate management [13]. In practice, this frequently manifests itself as excessive infiltration into soils, resulting in an excess of nutrients and, on occasion, their unintended infiltration into groundwater. Consequently, there is a growing interest annually not only in biogas facilities themselves but also in the processing of digestate, as the management of hydrated digestate gives rise to both logistical and legal issues [14]. For example, it is necessary to store the material for approximately six months in lagoons covered with a geomembrane or in special reinforced concrete tanks, since the decomposition processes are still ongoing [15]. Furthermore, the storage of hydrated digestate requires significant financial investment in the construction of reinforced concrete tanks or, in the case of lagoon construction, the acquisition of an additional land area of approximately 2 to 3 hectares [16]. A study [17] estimated that the costs associated with transporting hydrated digestate over distances greater than 10 km from a biogas plant, due to the significant amount of diesel fuel consumed, begin to exceed the profits associated with its sale as a fertilizer source. Therefore, it can be seen that a solution to the problem of high transport costs, or in the case of digestate storage, to the reduction in the volume of the tanks (lagoons), could be the separation of the digestate into a solid and a liquid fraction. It should be noted that the liquid fraction usually represents between 90 and 95% of the total digestate mass [18]. Therefore, it appears that the concept of digestate management could represent an opportunity for numerous small and medium enterprises that could be involved at this stage in the realization of municipal and agricultural biogas plants.

Given that digestate is treated not only as an alternative fertilizer but also as a source of water, it is essential that it undergoes a process of treatment that ensures that any contaminants present in the fermented biomass are not returned to the environment. In particular, ultrafiltration (UF) represents a potentially valuable technique in this context [19]. The process involves the retention of fine suspended solids, colloids, bacteria, and viruses. The transport mechanism is sieve-like, whereby particles larger than the diameter of the pores are excluded from passing through the membrane. The transmembrane pressure employed falls in the range of 0.1 to 1.0 MPa. The membranes employed in the UF process can be constructed from organic or inorganic materials. Organic membranes include polymeric membranes, comprising materials such as polysulfone, polyethersulfone, polyamide, and cellulosic materials. These membranes have been used in a multitude of applications in a range of fields [20,21]. However, their primary limitation is their limited chemical stability in aggressive systems, including aqueous solutions with low or high pH and solutions containing organic solvents. Nevertheless, they are distinguished by their ease of material processing, relatively low cost, and extensive range of properties. An alternative to polymeric membranes is ceramic membranes derived from inorganic materials. Inorganic membranes are distinguished from their organic counterparts by their high thermal, chemical, and mechanical resistance, as well as their capacity to withstand prolonged use [22]. Ceramic membranes produced from metal oxides, including Al, Ti, and Zr, have been implemented in the widest range of applications. Asymmetric ceramic membranes are made up of multiple layers. The support layer has a thickness of a few millimeters and comprises pores with a diameter of 1–10 μm. The intermediate layer, which has a thickness of 10–100 μm, comprises pores of a larger size, with a diameter of 50–100 μm. The resolution layer, which is the epidermal layer, is the thinnest, with a thickness of approximately 1 μm, and comprises the smallest pores, with a diameter in the range of 2 to 50 nm [23].

The utilization of membrane processes, including ultrafiltration (UF), for the purification of liquid digestate is expected to become increasingly common in the near future. This is due to the enhancement of membrane performance and the reduction in the costs of both membranes and membrane plants. At present, there is a paucity of literature on this subject, with only a small number of reports [24–26] available. Moreover, to the best of my knowledge, there are no published studies examining the applications of the aforementioned integrated process configuration for the treatment of digestates derived from diverse sources. The high inconvenience of digestate management, irrespective of its origin, has led to a growing interest in not only improving waste treatment technologies in biogas plants but also in developing new methods of digestate management and in enhancing existing techniques. A study was, therefore, conducted to compare the transport and separation properties of the tested membranes, which were made of organic and inorganic materials, in both a standalone membrane filtration process and an integrated process combining conventional treatment processes with an ultrafiltration process used to treat the liquid digestate fraction from different sources.

In consideration of the presented information, the comparison between the properties of the various membranes is original and has the potential to be highly beneficial in the selection of an appropriate method for the purification of the digestate.

2. Materials and Methods

This study employed two distinct types of liquid digest fraction, originating from a municipal biogas plant and an agricultural biogas plant, respectively. The first sample was obtained from a biogas plant processing the organic fraction of municipal waste located in the Lower Silesia Province (Poland, 50°53′15.5″ N 17°23′28.0″ E). The aforementioned fraction was separated from the digestate pulp by means of sedimentation centrifuges. The second fraction was obtained from an agricultural biogas plant located within the Silesian Voivodeship (Poland, 50°17′28.7″ N 18°32′50.3″ E). The samples were taken from a specific point of collection at the outlet of the anaerobic digestion (AD) reactor, prior to the mechanical dewatering installation. The characteristics of the two test solutions are presented in Table 1. The physicochemical analysis of the solutions was conducted in accordance with the Standard Methods for the Examination of Water and Wastewater, 23rd edition [27]. The details of the analytical procedures conducted and the instrumentation used are presented in Table 2.

Table 1. Composition of liquid digestate fraction from municipal and agricultural biogas plants.

	Liquid Digestate Fraction from the Municipal Waste Biogas Plant	Liquid Digestate Fraction from the Agricultural Biogas Plant
pH	7.17	7.21
Conductivity, mS/cm	25.21	14.95
Total suspended solids, mg/dm^3	670	3950
Chemical oxygen demand (COD), mg O_2/dm^3	29,360	38,595
5-day biochemical oxygen demand (BOD_5), mg O_2/dm^3	8690	12,320
Dissolved organic carbon (DOC), mg C/dm^3	8650	23,070
Na, mg/dm^3	487.2	521.3
K, mg/dm^3	1678.4	1966.5
Ca, mg/dm^3	89.2	104.7
Mg, mg/dm^3	672.2	101.9
Fe, mg/dm^3	6.2	15.9
Mn, mg/dm^3	4.4	1.5
Cu, mg/dm^3	0.230	0.545

Table 1. *Cont.*

	Liquid Digestate Fraction from the Municipal Waste Biogas Plant	Liquid Digestate Fraction from the Agricultural Biogas Plant
Zn, mg/dm^3	1.434	3.977
Hg, mg/dm^3	0.0040	0.0029
Co, mg/dm^3	0.156	0.069
Ni, mg/dm^3	0.320	0.147

Table 2. The analytical procedures conducted and the instrumentation employed.

Examined Parameter	Method	Apparatus
pH	Potentiometric method	Digital multimeter HQ40D with IntelliCAL™ PHC 101 electrode (Hach, Ames, IA, USA)
Conductivity	Conductometric method	
Total suspended solids	Weight-based method	-
COD	Bichromate method	
BOD_5	Dilution method	
DOC	NPOC high temperature oxidation method; thermal method	Hach IL550 carbon analyzer (Hach, Ames, IA, USA)
Na, K	Ion chromatography method	Thermo Scientific Dionex Aquion ion chromatograph with a conductometric detector for anions or cations analysis (Thermo Fisher Scientific, Waltham, MA, USA)
Ca, Mg	Titration method	-
Fe, Mn	Spectrophotometric method	Shimadzu UV-VIS 1800 (Shimadzu Corporation, Kyoto, Japan)
Cu, Zn, Co, Ni	Atomic absorption spectroscopy (ASA) with flame atomization	Atomic absorption spectrometer iCE 3500 (Thermo Fisher Scientific, Waltham, MA, USA)
Hg	Atomic absorption spectroscopy (ASA)—selective for Hg with concentration by amalgamation	AMA 254 mercury analyzer (Leco Corporation, St. Joseph, MI, USA)

The UF flat membranes used in the study were four polymeric membranes from Mann + Hummel Water & Fluid Solutions and four ceramic membranes from Tami Industries. The properties of the membranes used are presented in Table 3.

Table 3. Characteristics of the membranes used in the experiments [28,29].

Membrane Type		Membrane Material	Cut-Off	Max. Press., MPa	Max Temp., °C	pH Range	Active Filtration Area, cm^2
			Flat polymeric membranes				
PES 10 kDa	UF	Polyethersulfone	10 kDa	-	95	1–14	38.5
PES 30 kDa	UF		30 kDa				
C 10 kDa	UF	Regenerated cellulose	10 kDa		55	1–11	
C 30 kDa	UF		30 kDa				
			Flat ceramic membranes				
Ceram 1 kDa	FINE UF	TiO_2	1 kDa	0.4	350	2–14	56
Ceram 5 kDa	FINE UF		5 kDa				
Ceram 15 kDa	UF	ZrO_2	15 kDa			0–14	
Ceram 50 kDa	UF		50 kDa				

The tested membranes were distinguished by a distinctive asymmetric structure, comprising a thin epidermal layer and a thicker support layer. Figure 1 depicts microscopic images of surfaces of selected membranes.

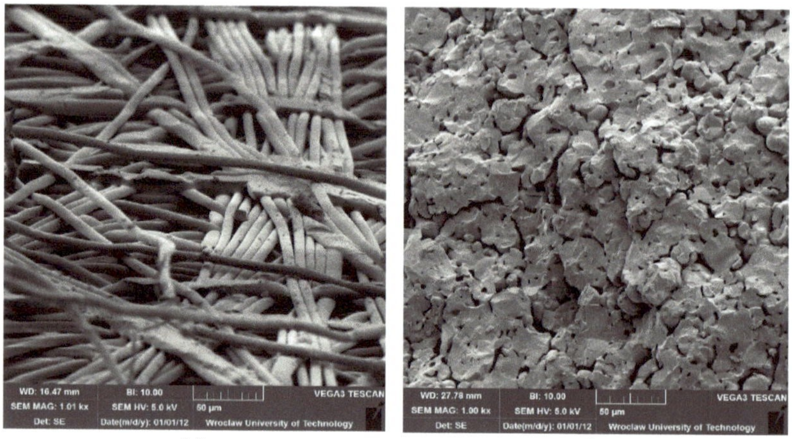

(**a**) (**b**)

Figure 1. SEM images of selected membranes (1000× magnification): (**a**) polymeric regenerated cellulose 10 kDa, (**b**) ceramic 5 kDa.

The pressure membrane filtration process was carried out using two installations, each equipped with a different type of membrane. The digestate was subjected to a 72 h sedimentation period prior to testing. The efficacy of the treatment of the liquid fraction of the digestate using polymeric membranes was evaluated through studies conducted on a test rig equipped with an Amicon 8400 chamber from Millipore (Figure 2). This chamber is designed to accommodate a dead-end filtration process and is intended for use with flat membranes. The Amicon 8400 chamber has a volume of 400 cm^3 and a membrane diameter of 76 mm. The UF chamber was positioned on an ARE magnetic stirrer from OMC Envag to guarantee that the concentrations were uniform throughout the solution volume. Additionally, experiments using ceramic membranes were conducted using a Sterlitech laboratory apparatus, consisting of a 316 SS pressure chamber with a capacity of 3.8 dm^3 (Figure 3). This apparatus is designed for operation with flat ceramic membranes in a dead-end configuration. The transmembrane pressure used in the trials was 0.3 MPa. All experiments were performed in duplicate to confirm the obtained results.

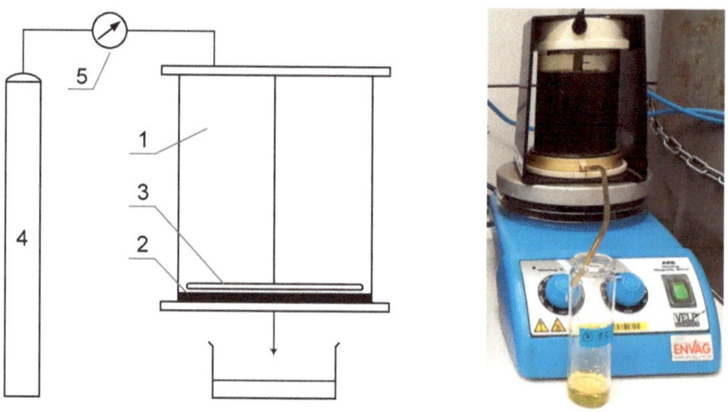

Figure 2. Laboratory setup with Millipore's Amicon 8400 chamber (1—Amicon 8400 chamber, 2—membrane, 3—stirrer, 4—compressed nitrogen cylinder, 5—regulator).

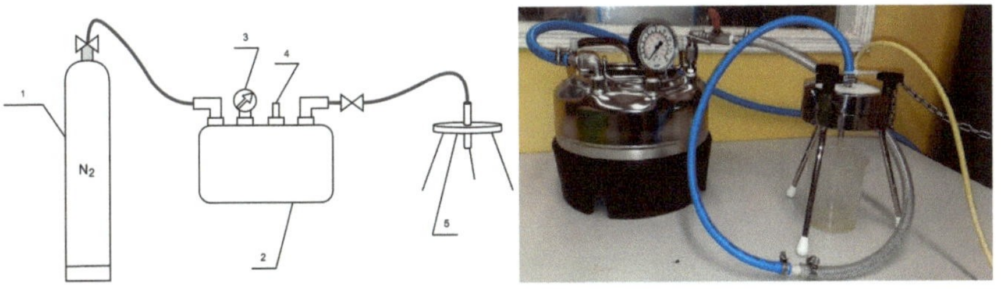

Figure 3. Sterlitech laboratory installation (1—compressed nitrogen cylinder, 2—pressure vessel, 3—manometer, 4—safety valve, 5—ceramic diaphragm in housing).

Studies on the purification of liquid digestate in an integrated process, which is a combination of conventional purification processes and membrane filtration, were carried out using the installations/test sites described above. Raw digestate was subjected to 72 h sedimentation and purification by filtration using a 1 µm pore size bag filter made of polyester felt material from ChemTech (Poland). The liquid was then coagulated with $FeCl_3 \cdots 6\,H_2O$ (Avantor Performance Materials Poland S.A., Gliwice, Poland) at 10 g/dm^3. The coagulation process was carried out by rapid stirring (150 rpm) for 2 min, then the stirrer speed was reduced to about 20 rpm and slow stirring was carried out for 20 min. Standalone coagulation lowered COD, BOD_5, and DOC to 20,640 g O_2/m^3, 6950 g O_2/m^3, and 6820 g C/m^3 for municipal digestate and 28,260 g O_2/m^3, 10,370 g O_2/m^3, and 17,240 g C/m^3 for agricultural digestate. The samples were then subjected to sedimentation and filtration using medium tissue filters for 30 min. The resulting clarified liquid was fed to both polymer and ceramic UF membranes.

Before carrying out the membrane filtration process, the polymeric membranes to be tested were subjected to conditioning, which consisted of filtering redistilled water through the membranes successively at different transmembrane pressures ranging from 0.1 to 0.4 MPa. The treatment was carried out until constant flux values were obtained.

The ceramic membranes, on the other hand, underwent a procedure to prepare them for proper operation before being used in the tests. This included alkaline cleaning by placing the membranes in NaOH solution (15–20 g/dm^3) at 80 °C for 30 min, rinsing until neutral pH, then acid cleaning and rinsing again until neutral pH and initial water flux was restored.

After each experiment, both polymeric and ceramic membranes were cleaned (chemically regenerated) with 0.1 mol/dm^3 NaOH solution (Avantor Performance Materials Poland S.A., Poland) and washed with redistilled water until the initial permeate flux values were obtained.

The transport properties of membranes used in UF were assessed by determining the permeate flux (J). Volumetric permeate flux is the volume of permeate obtained from a unit membrane surface area per unit time in a membrane process:

$$J = \frac{V}{A \cdot t}, \frac{m^3}{m^2 \cdot d} \quad (1)$$

where

- V—volume of permeate, m^3;
- A—membrane surface area, m^2;
- t—filtration time, d.

Relative membrane permeability values (J/J_0), calculated as the permeate flux (J) divided by the redistilled water flux of the new membrane (J_0), were used to characterize membrane fouling intensity.

The selectivity of the UF process was determined by the retention factor of the substance contained in the feed, also known as the reduction/removal ratio (R), according to the following formula:

$$R = \left(1 - \frac{c_p}{c_n}\right) \cdot 100, \% \quad (2)$$

where

- c_p—concentration of impurities in the treated solution, g/m^3;
- c_n—initial concentration of impurities in the solution to be purified, g/m^3.

The values R > 90% were determined with an error of less than 1%.

3. Results

3.1. Transport Properties of Membranes

When deciding on the suitability of individual membranes for the purification of the test solution, attention should be paid not only to their separation properties but also to their transport properties. The results of the redistilled water flux measurements for the organic and inorganic membranes tested are shown in Figure 4. They show that the hydraulic performance of a membrane depends significantly on the cut-off and the membrane material. For all membranes tested, the redistilled water flux increased with increasing cut-off values. Of all the polymeric membranes tested, the regenerated cellulose membrane with a cut-off of 30 kDa had the highest permeability (605.8 LMH). For a membrane with the same cut-off, but made of polyethersulfone, the permeate flux was 513.8 LMH. Of the ceramic membranes tested, the highest permeate flux, 48.3 LMH, was observed for the 50 kDa membrane. From the results obtained, it can be concluded that the hydraulic permeability of the polymeric membranes was much higher than that of the ceramic membranes.

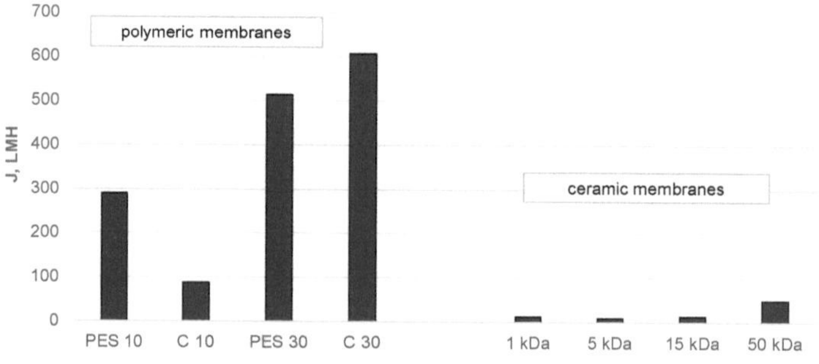

Figure 4. Redistilled water flux for polymeric and ceramic membranes.

A similar relationship to that for redistilled water was observed for all membranes used to filter the liquid fraction of both municipal and agricultural digestate (Figure 5)—membranes with higher cut-offs achieved higher permeate fluxes. This trend was observed regardless of whether a standalone UF process was used or whether it was preceded by pretreatment. The permeate flux values for both digestion fluids were also observed to be lower than those for redistilled water. This was due to an increase in flow resistance values due to membrane fouling. It was also observed that the greater the cut-off value of the membranes, the greater the decrease in permeate flux compared to redistilled water, indicating that internal fouling was the dominant factor in their case, caused by the penetration of impurities into the membrane pores.

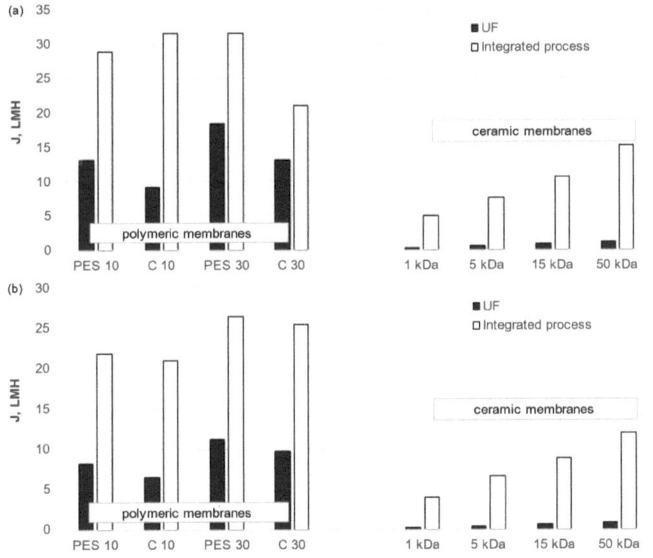

Figure 5. Effect of membrane type and cut-off value on permeate flux for the liquid fraction of municipal (**a**) and agricultural (**b**) digestate.

Furthermore, the content of the treated solution was found to exert a considerable influence on the membrane permeability. Filtration (either standalone or preceded by pretreatment) of the liquid fraction of agricultural digestate resulted in permeate fluxes that were inferior to those observed for the liquid fraction of municipal digestate. This indicates that contaminants present in agricultural digestate, by inducing alterations in the spatial configuration of organic matter and modifying its chemical properties, may intensify the membrane fouling phenomenon.

The membranes tested exhibited differences not only in their absolute hydraulic capacity but also in their susceptibility to fouling. Membrane fouling represents a significant challenge in the operation of membrane systems. It arises from the deposition of contaminants from the treated solution on the membrane surface and within the membrane structure. This phenomenon leads to a reduction in the permeate flux (in processes conducted at constant transmembrane pressure) or an increase in pressure (in processes conducted at constant permeate flux). As illustrated in Figure 6, the comparative permeability data (J/J_0) for the two digestate fluids under examination unambiguously demonstrate that the ceramic membranes exhibited a greater propensity for fouling than the polymeric membranes, particularly those with a cut-off of 1 and 50 kDa. This is likely attributable to their increased hydrophobicity, which results in a more pronounced reduction in permeability. The J/J_0 values of ceramic membranes with a cut-off of 5 kDa and 15 kDa are comparable to those obtained using regenerated cellulose membranes with a cut-off of 10 kDa and polyethersulfone membranes with a cut-off of 30 kDa, with a value of approximately 0.06. The membrane exhibiting the least degree of blockage was identified as an organic membrane of regenerated cellulose with a cut-off value of 30 kDa ($J/J_0 \cong 0.07$). This can be attributed to the membrane's notable hydrophilicity, which may have diminished the impact of the physicochemical characteristics of the filtered digestate on membrane fouling. The use of hydrophilic membranes serves to diminish the accumulation of material at the solution–membrane interface and to extend the duty cycle of the membrane. The outcomes align with the documented wetting angles of the membranes (Table 3), substantiating the hydrophilic or hydrophobic nature of the membranes. Membranes with an identical cut-off value, yet manufactured from disparate materials, also exhibit variation in their average pore size, which exerts an influence on the extent of membrane fouling. Irrespective of the membrane type subjected to examination, the resulting agglomerates of organic compounds formed a filter cake on the membrane surface. Similar findings were observed when comparing the fouling intensity of organic and inorganic membranes in an integrated process, which included sedimentation, filtration, coagulation, sedimentation, and UF. Subsequently, it was discovered that polymeric membranes exhibited a greater susceptibility to fouling than ceramic membranes. The J/J_0 values for the liquid fraction of municipal digestate were 0.10 and 0.06 for polymeric membranes with cut-offs of 10 kDa and 30 kDa, respectively. These values were observed for PES membranes, while for C membranes, the values were 0.36 and 0.03, respectively. In contrast, the values were significantly higher for ceramic membranes, with the highest value observed in membranes with a cut-off of 5 kDa (0.85). For the liquid fraction of agricultural digestate, the corresponding values were 0.07 and 0.05 for PES membranes, 0.24 and 0.04 for C membranes, and 0.74 for the 5 kDa ceramic membrane, respectively. The findings indicate that the utilization of a pretreatment process involving $FeCl_3 \cdots 6\,H_2O$ had a pronounced impact on the extent of membrane fouling. The purification of the liquid fraction of fermentation in the integrated process, as analyzed in all cases, resulted in a notable reduction in the intensity of membrane fouling. The preceding pretreatment of the solution, thus, results in a notable reduction in the fouling of the membranes under analysis. This was particularly evident when sedimentation, filtration, coagulation, and sedimentation processes were combined in a sequence using a ceramic membrane with a cut-off of 5 or 15 kDa. The results suggest that, in the case of more compact membranes, the use of coagulation preceded by sedimentation and filtration allows the removal of compounds from the digestate that, in the absence of pretreatment, would otherwise settle on the membrane surface and block it. The results obtained for a membrane with a cut-off of 1 kDa represent an exception to this rule. The aforementioned filtration processes on other ceramic membranes also improve the transport properties of the membranes, although to a lesser extent. This may be attributed to the penetration of residual fractions in solution into the membrane pores and their subsequent blocking.

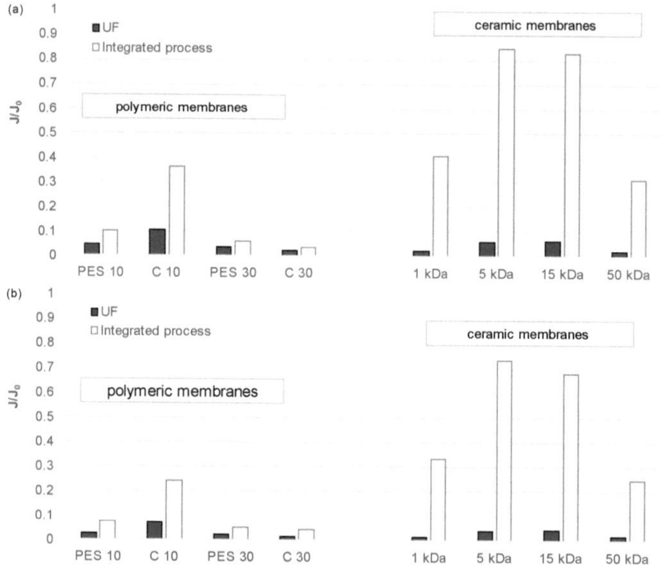

Figure 6. Relative permeability during standalone UF and during the integrated process: sedimentation/filtration/coagulation/sedimentation/UF of (**a**) the liquid fraction of municipal digestate and (**b**) the liquid fraction of agricultural digestate using polymeric and ceramic membranes.

Furthermore, a comparison was conducted between the fouling intensities of three UF membranes: one ceramic membrane with a cut-off of 15 kDa and two polymeric membranes with a comparable cut-off value of 10 kDa, under conditions of altered pH of the filtered solution (Figure 7). It was observed that an increase in pH value resulted in an increase in the intensity of membrane fouling for both test solutions, including those comprising organic membranes made of polyethersulfone and inorganic membranes. On the contrary, no change in the intensity of membrane fouling was observed for regenerated cellulose membranes. It can, therefore, be concluded that the use of membranes made of highly hydrophilic materials, such as regenerated cellulose, can, to some extent, reduce the influence of the physicochemical properties of the filtered solution on membrane fouling.

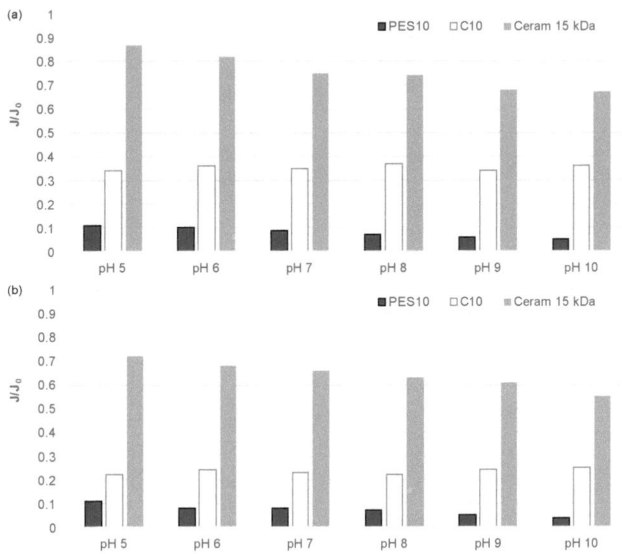

Figure 7. pH dependence of the relative permeability of polymeric and ceramic membranes for integrated process: sedimentation/filtration/coagulation/sedimentation/UF of the liquid fraction of (**a**) municipal and (**b**) agricultural digestates.

3.2. Separation Properties of Membranes

In order to evaluate the efficacy of the membranes under investigation, the impact of membrane type on the separation efficiency of organic compounds from both solutions was subjected to analysis. A comparison of the purification efficiencies of the UF process and the integrated process combining conventional purification processes with the UF process on organic and inorganic membranes is presented in Figures 8 and 9. The cut-off value stated by the manufacturer is often a determining factor in the selection of the appropriate membrane. Therefore, it was deemed appropriate to refer to this value when comparing the two membrane types. The results of the analysis indicate that the removal efficiency of organic macromolecules was markedly higher for ceramic membranes than for polymeric membranes with comparable cut-off values. To illustrate, the use of a ceramic membrane with a cut-off value of 15 kDa in a standalone UF process for the treatment of municipal digestate resulted in retention rates of 35%, 34%, and 29% for DOC, BOD_5, and COD, respectively. This equates to a removal of 2910 to 1880 g C/m^3 and a reduction from 1370 to 1110 g O_2/m^3. The PES 10 membrane demonstrated comparable performance at 25, 28, and 11% (reduction to 2190 g C/m^3, 1370 g O_2/m^3, and 5220 g O_2/m^3, respectively). The results demonstrated that the PES 10 membrane exhibited the highest removal efficiency at 26, 28, and 11% (2160 g C/m^3 and 5250 g O_2/m^3, respectively), while the C 10 membrane demonstrated the lowest removal efficiency at the same concentrations (1385 g O_2/m^3 and

5220 g O_2/m^3, respectively). It can, thus, be concluded that the material of the polymeric membrane employed, irrespective of the process utilized, had no discernible impact on the separation properties of the membranes under examination. Both polyethersulfone and regenerated cellulose yielded permeates with comparable organic content.

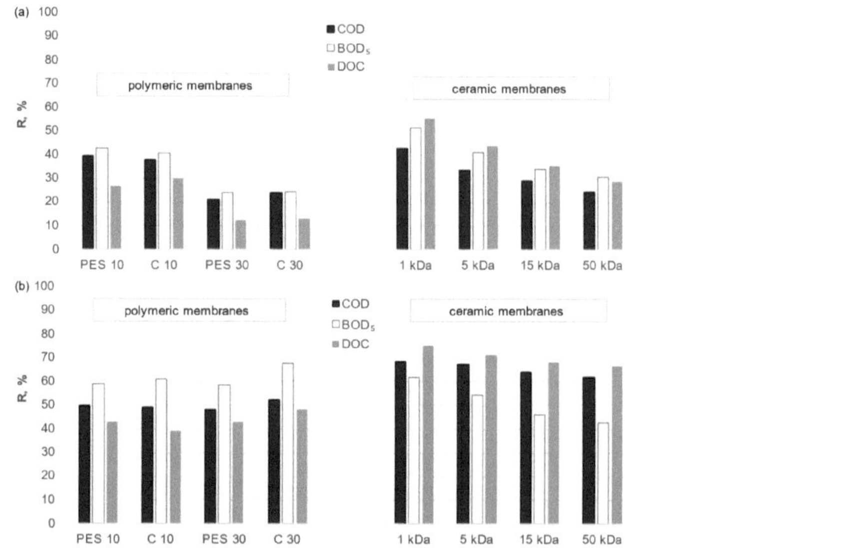

Figure 8. Effectiveness of DOC removal and reduction in BOD_5 and COD when conducting (**a**) UF and (**b**) integrated process: sedimentation/filtration/coagulation/sedimentation/UF of the liquid fraction of municipal digestate using polymeric and ceramic membranes.

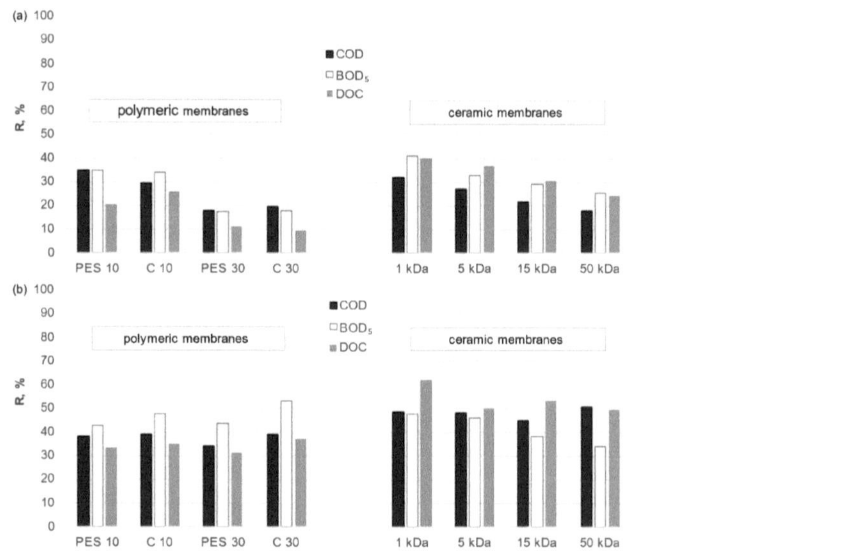

Figure 9. Effectiveness of DOC removal and reduction in BOD_5 and COD when conducting (**a**) UF and (**b**) integrated process: sedimentation/filtration/coagulation/sedimentation/UF of the liquid fraction of agricultural digestate using polymeric and ceramic membranes.

The ceramic membrane with a cut-off of 1 kDa yielded the most favorable results in the standalone membrane separation process. The separation efficiency for liquid municipal digestate was DOC at 55%, BOD_5 at 51%, and COD at 43% (reduction to 1300 g C/m^3, 930, and 3365 g O_2/m^3, respectively). On the contrary, the results for liquid agricultural digestate were 40%, 41%, and 32% (reduction to 13,830 g C/m^3, 7265, and 26,200 g O_2/m^3, respectively). The contaminant removal of all membranes analyzed was determined by the sieve mechanism. The ratio of contaminant size to membrane pore diameter had a significant effect on removal efficiency, as discussed in [30].

Similar outcomes were observed when membrane separation was coupled with conventional treatment processes. The retention factors for DOC, BOD_5, and COD were higher when ceramic membranes were employed than when polymeric membranes were used, irrespective of the nature of the solution undergoing treatment.

In both the standalone UF process and the integrated process, membrane cut-off resolution emerged as a significant consideration. As the cut-off point and, consequently, the average pore size increased, the quality of the treated solution's permeate declined. The utilization of membranes with a more compact structure enabled the attainment of permeate with a reduced organic content. Furthermore, the sedimentation process conducted at the pretreatment stage prior to coagulation facilitated the partial clarification of the solution by reducing the concentration of larger particles in the analyzed solution. In addition, the suspended fraction was eliminated by filtration. Therefore, the final quality of the investigated digestate was determined by the combination of the efficiency of the conventional purification processes applied and the separation properties of the membrane. The separation mechanism in the analyzed integrated process is a combination of the coagulation effect of $FeCl_3 \cdots 6\,H_2O$ of the presuspended solution and the sieve mechanism on the membranes tested.

The specific type of liquid digestate fraction subjected to analysis is also of considerable importance. It has been demonstrated that the efficiency of the separation of organic pollutants from liquid digestate is significantly influenced by the composition of the treated solution, regardless of whether the membrane is organic or inorganic. Despite the comparable content of organic compounds in the liquid fraction of agricultural-derived digestate and the liquid fraction of municipal-derived digestate, the former did not contain the same amount of other organic and inorganic pollutants. It was found that the retention rates of DOC, BOD_5, and COD during the treatment of municipal digestate were significantly higher than those observed for agricultural digestate. This trend was observed for all membranes tested. The presence of inorganic ions in liquid agricultural digestate has been shown to influence both the spatial configuration of organic particles and the surface charge, which can facilitate the transport of organic substances through the membrane. This highlights the need for further research and verification of the results obtained for different types of liquid digestate fractions.

As was the case in the determination of the transport properties of the membranes, a comparison was made between the effect of the solution reaction on the separation properties of polymeric and ceramic membranes with comparable cut-off values (Figures 10 and 11). The results of the analysis demonstrate a notable influence of the pH of the test solution on the purification of the liquid fraction of municipal and agricultural digestate on both types of membrane, regardless of the purification process used. An increase in the pH value within the range of 5 to 10 was observed to result in increased effectiveness of the reduction in DOC for both organic and inorganic membranes. The lowest retention was observed at pH 5, while an increase in the pH value in the analyzed range resulted in an improvement in the separation efficiency of organic macromolecules for both organic and inorganic membranes. This effect is attributed to the fact that at pH = 5, organic macromolecules are poorly dissociated, resulting in a relatively small spatial dimension and enhanced penetration through the membranes. As the pH value of the solution increases, the phenolic and carboxyl functional groups of organic compounds dissociate, resulting in their larger spatial dimension and easier retention on the membrane. Furthermore, the adsorption of anions from the solution onto the pore walls of the membrane is also possible [31], which results in a reduction in the membrane pore diameter and stronger electrostatic interactions between organic macroanions and the membrane.

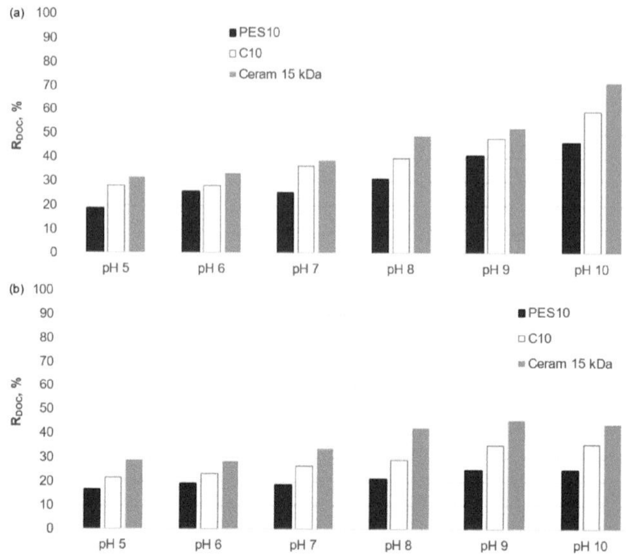

Figure 10. Influence of solution pH on the degree of DOC reduction on organic and inorganic membranes during treatment of the liquid fraction of municipal digestate by (**a**) UF and (**b**) integrated: sedimentation/filtration/coagulation/sedimentation/UF processes.

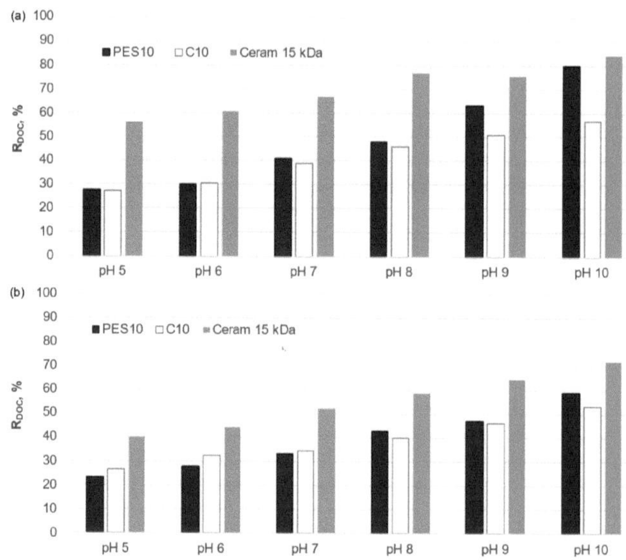

Figure 11. Influence of solution pH on the degree of DOC reduction on organic and inorganic membranes during treatment of the liquid fraction of agricultural digestate by (**a**) UF and (**b**) integrated: sedimentation/filtration/coagulation/sedimentation/UF processes.

4. Conclusions

The suitability of the UF process, either alone or in combination with conventional processes to form an integrated process, has been confirmed by studies for the treatment of different types of liquid digestate (from an agricultural biogas plant and from a biogas plant treating the organic fraction of municipal waste). Given the fundamental differences between the two types of digestate, it was imperative to conduct dedicated research on each. The findings demonstrate that the treatment of liquid digestate in the aforementioned manner facilitates the recovery of water, which, following purification, can be utilized in agricultural applications, including the preparation of fertilizers or irrigation. A comparative analysis of the transport and separation properties of membranes comprising organic and inorganic materials employed for the purification of liquid digestate yielded the following conclusions:

1. Polymeric membranes are characterized by higher hydraulic permeability values compared to ceramic membranes.
2. Permeate fluxes during membrane filtration (standalone or preceded by pretreatment) of the liquid fraction of agricultural digestate were lower than those measured for the liquid fraction of municipal digestate.
3. Inorganic membranes are more susceptible to fouling when used in a standalone ultrafiltration (UF) process, whereas organic membranes are more susceptible in an integrated process.
4. The removal efficiency of organic macromolecules was significantly higher when using ceramic membranes than polymeric membranes with a comparable cut-off value.
5. For both types of membranes tested, an increase in the pH value of the solution resulted in an increase in the intensity of membrane fouling, as well as an improvement in the separation efficiency of organic macromolecules.

Funding: This research received no external funding.

Institutional Review Board Statement: Not applicable.

Data Availability Statement: The data that support the findings of this study are available from the corresponding author upon reasonable request.

Conflicts of Interest: The author declare no conflicts of interest.

References

1. Grafton, R.Q.; Williams, J.; Jiang, Q. Food and Water Gaps to 2050: Preliminary Results from the Global Food and Water System (GFWS) Platform. *Food Secur.* **2015**, *7*, 209–220. [CrossRef]
2. Steffen, W.; Richardson, K.; Rockström, J.; Cornell, S.E.; Fetzer, I.; Bennett, E.M.; Biggs, R.; Carpenter, S.R.; de Vries, W.; de Wit, C.A.; et al. Planetary Boundaries: Guiding Human Development on a Changing Planet. *Science* **2015**, *347*, 1259855. [CrossRef]
3. Hertwich, E.; Lifset, R.; Pauliuk, S.; Heeren, N. *Material Efficiency Strategies for a Low-Carbon Future Summary for Policymakers*; UN Environment Programme: Cambridge, UK, 2020.
4. Franco-García, M.-L.; Carpio-Aguilar, J.C.; Bressers, H. *Towards Zero Waste, Circular Economy Boost: Waste to Resources*; Springer: Berlin/Heidelberg, Germany, 2019; pp. 1–8.
5. Ma, H.; Guo, Y.; Qin, Y.; Li, Y.-Y. Nutrient Recovery Technologies Integrated with Energy Recovery by Waste Biomass Anaerobic Digestion. *Bioresour. Technol.* **2018**, *269*, 520–531. [CrossRef]
6. Holm-Nielsen, J.B.; Al Seadi, T.; Oleskowicz-Popiel, P. The Future of Anaerobic Digestion and Biogas Utilization. *Bioresour. Technol.* **2009**, *100*, 5478–5484. [CrossRef]
7. Drosg, B.; Fuchs, W.; Al Seadi, T.; Madsen, M.; Linke, B. *Nutrient Recovery by Biogas Digestate Processing*; Tamarind: London, UK, 2015.
8. Sikora, J.; Sadowska, U.; Klimek-Kopyra, A.; Gliniak, M. The Impact of Biochar Dosing with Simultaneous Fertilization with a Biogas Plant Digestate on the Volumetric Density of Soil. *E3S Web Conf.* **2019**, *132*, 02006. [CrossRef]
9. Kumar, A.; Samadder, S.R. Performance Evaluation of Anaerobic Digestion Technology for Energy Recovery from Organic Fraction of Municipal Solid Waste: A Review. *Energy* **2020**, *197*, 117253. [CrossRef]
10. Hosseini Koupaie, E.; Azizi, A.; Bazyar Lakeh, A.A.; Hafez, H.; Elbeshbishy, E. Comparison of Liquid and Dewatered Digestate as Inoculum for Anaerobic Digestion of Organic Solid Wastes. *Waste Manag.* **2019**, *87*, 228–236. [CrossRef]

11. Carlsson, M.; Lagerkvist, A.; Morgan-Sagastume, F. The Effects of Substrate Pre-Treatment on Anaerobic Digestion Systems: A Review. *Waste Manag.* **2012**, *32*, 1634–1650. [CrossRef]
12. Kovačić, Đ.; Lončarić, Z.; Jović, J.; Samac, D.; Popović, B.; Tišma, M. Digestate Management and Processing Practices: A Review. *Appl. Sci.* **2022**, *12*, 9216. [CrossRef]
13. Baştabak, B.; Koçar, G. A Review of the Biogas Digestate in Agricultural Framework. *J. Mater. Cycles Waste Manag.* **2020**, *22*, 1318–1327. [CrossRef]
14. Nowak, M.; Czekała, W. Sustainable Use of Digestate from Biogas Plants: Separation of Raw Digestate and Liquid Fraction Processing. *Sustainability* **2024**, *16*, 5461. [CrossRef]
15. Makara, A.; Kowalski, Z.; Fela, K. Disposal of After-Fermentation Substance in the Aspect of Ecological Safety. Prace Naukowe Akademii im. Jana Długosza w Częstochowie. *Tech. Inform. Inżynieria Bezpieczeństwa* **2017**, *5*, 177–190. [CrossRef]
16. Żakowicz, A.; Zimiński, K.; Obolewicz, J. Analysis of Structural Solutions of Reinforced Concrete Tanks Used in Agricultural Farms. *Inżynieria Bezpieczeństwa Obiektów Antropog.* **2023**, *1*, 22–32. [CrossRef]
17. Delzeit, R.; Kellner, U. The Impact of Plant Size and Location on Profitability of Biogas Plants in Germany under Consideration of Processing Digestates. *Biomass Bioenergy* **2013**, *52*, 43–53. [CrossRef]
18. Kasprzycka, A.; Lalak, J.; Tys, J.; Chmielewska, M. Selected Methods for Management of Post-Fermentation Sediment. In *The Use of Extrusion Processing in Digested Sludge Management (A Review)*; Acta Agrophysica: Lublin, Poland, 2016; Volume 23.
19. Yue, C.; Dong, H.; Chen, Y.; Shang, B.; Wang, Y.; Wang, S.; Zhu, Z. Direct Purification of Digestate Using Ultrafiltration Membranes: Influence of Pore Size on Filtration Behavior and Fouling Characteristics. *Membranes* **2021**, *11*, 179. [CrossRef]
20. Baker, R.W. Ultrafiltration. In *Membrane Technology and Applications*; Wiley: Hoboken, NJ, USA, 2023; pp. 241–286.
21. Li, X.; Jiang, L.; Li, H. Application of Ultrafiltration Technology in Water Treatment. *IOP Conf. Ser. Earth Environ. Sci.* **2018**, *186*, 012009. [CrossRef]
22. Zhang, Y.; Tan, Y.; Sun, R.; Zhang, W. Preparation of Ceramic Membranes and Their Application in Wastewater and Water Treatment. *Water* **2023**, *15*, 3344. [CrossRef]
23. He, Z.; Lyu, Z.; Gu, Q.; Zhang, L.; Wang, J. Ceramic-Based Membranes for Water and Wastewater Treatment. *Colloids Surf. A Physicochem. Eng. Asp.* **2019**, *578*, 123513. [CrossRef]
24. Gerardo, M.L.; Aljohani, N.H.M.; Oatley-Radcliffe, D.L.; Lovitt, R.W. Moving towards Sustainable Resources: Recovery and Fractionation of Nutrients from Dairy Manure Digestate Using Membranes. *Water Res.* **2015**, *80*, 80–89. [CrossRef]
25. Chiumenti, R.; Chiumenti, A.; Da Borso, F. Digestate Treatment by Means of a Full Scale Membrane System: An Innovative Method for Managing Surplus Nitrogen and for Valorising Farm Effluents. In Proceedings of the 14th Ramiran International Conference, Lisbon, Portugal, 13–15 September 2010.
26. Camilleri-Rumbau, M.S.; Norddahl, B.; Wei, J.; Christensen, K.V.; Søtoft, L.F. Microfiltration and Ultrafiltration as a Post-Treatment of Biogas Plant Digestates for Producing Concentrated Fertilizers. *Desalination Water Treat* **2015**, *55*, 1639–1653. [CrossRef]
27. Baird, R.; Eaton, A.D.; Rice, E.W.; Bridgewater, L.; Federation, W.E. *Standard Methods for the Examination of Water and Wastewater*, 23rd ed.; American Public Health Association: Washington, DC, USA, 2017; ISBN 9780875532875.
28. MANN+HUMMEL. Water & Fluid Solutions Flat Sheet Membrane Data Sheets. Available online: https://water-membrane-solutions.mann-hummel.com/en/downloads.html (accessed on 12 August 2024).
29. Sterlitech. Ceramic Membrane Filters. Available online: https://www.sterlitech.com/media/wysiwyg/pdfs/Sterlitech_Catalog2016_Ceramic_.pdf (accessed on 14 June 2024).
30. Urbanowska, A.; Kabsch-Korbutowicz, M. Properties of Flat Ceramic Membranes and Their Application for Municipal Digestate Liquid Fraction Purification. *J. Membr. Sci. Res.* **2023**, *9*, 556692. [CrossRef]
31. Elimelech, M.; Chen, W.H.; Waypa, J.J. Measuring the Zeta (Electrokinetic) Potential of Reverse Osmosis Membranes by a Streaming Potential Analyzer. *Desalination* **1994**, *95*, 269–286. [CrossRef]

MDPI AG
Grosspeteranlage 5
4052 Basel
Switzerland
Tel.: +41 61 683 77 34

Membranes Editorial Office
E-mail: membranes@mdpi.com
www.mdpi.com/journal/membranes

www.ingramcontent.com/pod-product-compliance
Lightning Source LLC
Chambersburg PA
CBHW041345210926
51850CB00013B/2993
* 9 7 8 3 7 2 5 8 2 4 3 0 4 *